DISCARDED

GARLAND STUDIES ON

INDUSTRIAL PRODUCTIVITY

edited by
STUART BRUCHEY
UNIVERSITY OF MAINE

A GARLAND SERIES

COMPETITIVENESS IN UNITED STATES GRAIN EXPORTS

Cost-Effective Shipping Patterns in International Rice Marketing

MEI MIRANDA ZHANG

GARLAND PUBLISHING, INC.
NEW YORK & LONDON / 1996

Copyright © 1996 Mei Miranda Zhang
All rights reserved

Library of Congress Cataloging-in-Publication Data

Zhang, Mei Miranda, 1958–
  Competitiveness in United States grain exports : cost-effective shipping patterns in international rice marketing / Mei Miranda Zhang.
    p. cm. — (Garland studies on industrial productivity)
  Includes bibliographical references and index.
  ISBN 0-8153-2352-2 (alk. paper)
  1. Rice trade—United States. 2. Rice—United States—Transportation. 3. Shipping—United States. I. Title. II. Series.
HD9066.U45Z48 1996
382'.41318'0973—dc20   95-51129

Printed on acid-free, 250-year-life paper
Manufactured in the United States of America

TO *MILSON* AND *HANSON*

Contents

List of Tables ix

List of Illustrations xi

Preface xiii

I. Overview 3

    U.S. Rice Production and Exports 3
    Review of Literature 7
    U.S. Shipping Industry 9
    Flag of Registry 14
    The U.S. Cargo Preference Policies 18
    Rice Transportation Vessels 21
    Cargo Vessel Charters 24
    Terms of Shipments 26
    Summary 28

II. The Theoretical Framework 31

    International Trade 31
    Simple Models of Rice Trade 33
    Transportation Cost 36
    Effect of Distance, Ship Size and Trade Volume on
        Shipping Cost 41
    Role of Transportation Cost in International Trade 45
    Summary 49

III. Estimation of Shipping Cost 53

Shipping Cost and Measurement 53
Ports of Origin and Ports of Destination 55
Lay Days and Days at Sea 57
Assumptions of the Estimation 70
The Estimation of Shipping Cost 72
Summary 86

IV. The Transportation Models 101

Methodology and Data Requirements 101
Assumptions of the Model 107
Procedure of the Model 107
Analysis of the Results 108
Summary 128

V. Summary and Conclusions 137

Summary 137
Suggestions 142

Appendix A 145

Appendix B 147

Bibliography 165

Index 175

Tables

Table 1. Trend of Rice Production in Major Rice Exporting Countries   5
Table 2. Trend of Rice Export Price Index   6
Table 3. Trend of Rice Exports in Major Rice Exporting Countries   6
Table 4. U.S.-flag Privately Owned Merchant Fleet   11
Table 5. U.S. Dry-Bulk Carriers in International Trade   12
Table 6. U.S. Dry-Bulk Export and Import Trades   13
Table 7. The U.S. Effective-Control Fleet   15
Table 8. Monthly Average Compensation for Selected Ship Workers by Country   17
Table 9. Importance to Liner Vessels of Carriage of Government Preference Cargoes   19
Table 10. Share of Shipment of U.S. Southern Region Rice Exports by Different Vessels   22
Table 11. U.S. Southern Rice Export Regions and Base Points   56
Table 12. Countries and Regions Importing Rice from the U.S.   58
Table 13. Days in Port, Days at Sea, and Distance Between Export Points (for 25,000 and 30,000 DWT ships)   62
Table 14. Days in Port, Days at Sea, and Distance Between Export Points (for 35,000 DWT ships)   65
Table 15. Days in Port, Days at Sea, and Distance Between Export Points (for 40,000 DWT ships)   68
Table 16. Characteristics of Bulk Rice Vessels by Different Flag and Size   72
Table 17. Vessel Ownership Expenses: Interest and Depreciation   75
Table 18. Vessel Ownership Expenses: Return on Investment   77
Table 19. Five Other Vessel Ownership Expenses   78
Table 20. Total Vessel Ownership Expenses   79
Table 21. Port Expenses by Vessel Size and Flag   80
Table 22. Calculated Transportation Rate for Rice Exports (For 25,000 DWT Vessels)   87
Table 23. Calculated Transportation Rate for Rice Exports (For 30,000 DWT Vessels)   90

Table 24. Calculated Transportation Rate for Rice Exports (For 35,000 DWT Vessels) 93
Table 25. Calculated Transportation Rate for Rice Exports (For 40,000 DWT Vessels) 96
Table 26. Rice Surplus Exported from U.S. Southern Region 105
Table 27. Rice Deficit by World Region 106
Table 28. Actual Shipping Pattern for U.S. Southern Rice Exports 110
Table 29. Model I: Optimal Shipping Pattern for U.S. Rice Exports 115
Table 30. Model II: Optimal Shipping Pattern for U.S. Rice Exports 118
Table 31. Model III: Optimal Shipping Pattern for U.S. Rice Exports 121
Table 32. Model IV: Optimal Shipping Pattern for U.S. Rice Exports 124
Table 33. Model V: Optimal Shipping Pattern for U.S. Rice Exports 129
Table 34. Model VI: Optimal Shipping Pattern for U.S. Rice Exports 132

Table A.1. Calculated Transportation Rates for 30,000 DWT, Japan-flag and 40,000 DWT Liberia-flag Vessels (for Model II) 149
Table A.2. Calculated Transportation Rates for 30,000 DWT, Japan-flag and 40,000 DWT Liberia-flag Vessels (for Model III) 152
Table A.3. Adjusted Port Expenses for Port of Houston and Port of New Orleans (for Model IV) 155
Table A.4. Calculated Transportation Rates for Shipment from Port of New Orleans to the Selected Destinations (for Model IV) 156
Table A.5. Calculated Transportation Rates for Shipment from Port of Houston to the Selected Destinations (for Model IV) 157
Table A.6. Adjusted Total Vessel Ownership Expenses for U.S.-flag Ship (for Model V) 159
Table A.7. Calculated Transportation Rates for U.S.-flag Vessels Over Selected Routes (for Model V) 160
Table A.8. Adjusted Total Vessel Ownership Expenses for U.S.-flag Ship (for Model VI) 162
Table A.9. Calculated Transportation Rates for U.S.-flag Vessels Over Selected Routes (for Model VI) 163

Illustrations

Figure 1. Relative Costs for Crew for a Typical Containership Operating Under Various Flags   16
Figure 2. Free Trade Model for Rice Between Two Countries   35
Figure 3. Free Trade Model for Rice: Three-Country Case   37
Figure 4. The Effect of Transportation Cost on International Rice Market: Two-Country Case   47
Figure 5. The Effect of Transportation Cost on International Rice Market: Three-Country Case   48

Preface

The United States is one of the major grain producing and exporting countries in the world. The grain provides economic livelihood for many millions of farm families in the world and those engaged in marketing and distribution. Rice is a major crop for the United States in international grain trade though it is not a major crop for consumption. One of the questions the U.S. grain industry has been facing is the question of how to keep its appropriate share in the world market. Rice export, for example, is a major activity for the U.S. grain industry; however, as rice production has expanded in the U.S. and in foreign countries, the U.S. rice exports have been falling while in some other countries exports have increased rapidly over last two decades. There are several reasons for this decline in the U.S. rice exports, one of which is transportation cost. This cost affects the competitive position of the U.S. rice industry in the international rice market.

The southern region of the United States is a major rice producing and exporting area. The cost of shipping rice on various sizes of vessels from the southern region of the U.S. to international markets can be one of the major sources of trade restrictions for U.S. rice exports. Any change in transportation costs will have a great impact on rice exporters and others involved in the movement of rice from the U.S. to importing countries and thus will severely affect the competitiveness of U.S. rice industry in the world market. Since transportation cost of rice exports has a direct effect on rice prices, it is desirable that rice exports be shipped over an optimal, least-cost routing pattern. Such an optimal shipping pattern for U.S. rice exports can be found with the help of the transportation model. A study that analyzes the U.S. rice export situation, the cost of shipping rice to international markets and the optimal shipping patterns for U.S. rice exports is necessary to provide a better understanding of the impact of cost of transportation service rates on rice shipments from the U.S. to importing countries and the competitiveness of U.S. grain industry in the world market. This book examines the transportation cost and the role of this cost in international trade, different types of flag vessels, and the least-cost optimal exporting patterns for U.S. rice shipment.

The purpose of this book is to determine the cost per ton of shipping rice for selected sizes of bulk vessels from various U.S. southern ports of origin to specific foreign import ports. These cost data are then used in a transportation model to estimate a least-cost shipping pattern for U.S. rice exports.

For background information, the existing transportation system is identified, to the extent possible, by making personal contact and by using secondary data where available. These data includes rice exports from the U.S. southern region, shipment by type of vessel (liner, tramp, tanker), by type of flag (U.S., Japan, Liberia), imports of U.S. rice by various foreign countries, shipping costs and port charges or costs. Certain procedures are developed to determine the types of costs involved in the shipment of rice exports by different vessels. For example, the major types of costs of owning and operating a vessel can be classified as vessel ownership expenses, at-sea expenses and port expenses. The cost per ton of shipping rice between a port of origin and a port of destination is then estimated. The analysis of rice shipments is based on data of relevant technology and cost of rice shipping obtained from correspondence with port authorities, Maritime Administration, and other sources. Based on these data, several transportation models under different situations are developed to derive the kind of transportation system that could efficiently move U.S. rice to major world markets at least cost.

CHAPTER OUTLINE

An overview that provides some important background information is presented in Chapter I. Chapter II develops the theoretical framework concerning transportation cost and its role in international trade. Both chapters provide a necessary understanding of the importance of estimating cost of rice export shipment from the U.S. to world markets. Chapter III determines the types of costs involved in rice shipments. Also in that chapter specific measurements are developed for estimating and calculating shipping cost per ton between any two ports with different types of flags and different sizes of vessels. In Chapter IV, certain techniques are used to conduct transportation models using collected data and shipping cost estimated in Chapter III, through which the least-cost shipping patterns are determined. Comparisons and evaluations of the results generated from the models are also developed in Chapter IV to analyze the effects of different policies on the optimal transportation pattern. Finally, in Chapter V, summary, conclusions, implications and suggestions for further research are presented.

POTENTIAL BENEFIT

Maintaining the lowest possible transportation cost will enable the U.S. to be more competitive in international markets by lowering prices of the exports. Knowing how the current and alternative levels of transportation costs affect rice exports will enable producers, exporters, government and others to apply appropriate policies on rice production, transportation and marketing.

ACKNOWLEDGMENTS

The author wishes to express her indebtedness to those who gave their talents and time to assist in the preparation of this book. Special thanks go to *Albert J. Allen* of Mississippi State University, for his interest, guidance, and contribution to the completion of this book. Appreciation is also extended to *Harold A. Smith* for his meticulous editing of the manuscript. A debt of gratitude is owed to *Jeff Wilson* for his help in model development in the book. Acknowledgment also goes to *Eva Yang* for her patience and excellence in typing the manuscript.

Competitiveness in United States Grain Exports

I
Overview

This chapter provides necessary background information for the analysis of transportation cost. The information includes the situation of U.S. rice production and exports, literature review, U.S. shipping industry, types of flag and types of vessels. Terms of shipments and U.S. cargo preference policies are also provided in order to better examine transportation service rates in shipping rice from U.S. southern ports to world markets.

U.S. RICE PRODUCTION AND EXPORTS

Rice is a very important part of many countries' agricultural industry. As a commodity, rice is the major staple food for most developing countries and less developed countries. Rice is a major crop for the United States in international trade. The major rice producing areas are the Grand Prairie and Northeast areas of Arkansas, the Mississippi River Delta, Southeast Louisiana, the Coast Prairie of Texas, and the Sacramento Valley of California. For example, those four southern states, together with the state of California, accounted for more than 98 percent of the total rice production among all the states that have reported rice acreage in 1980s.[1]

Rice produced in the U.S. can be classified into three basic categories: long-grain rice (which is the most popular type), medium-grain rice and short-grain rice. Over the last decade, the three different types of rice accounted for 57 percent, 34 percent and 9 percent, respectively, of all U.S. rice production.[2] The U.S. southern region producers grow mostly long-grain and some medium-grain rice.

Rice is not a major crop for U.S. domestic consumption, which accounted for around 40 percent of total U.S. rice production. In 1984, the average per capita consumption of rice in the United States was 8.6 pounds.[2] The main domestic uses of rice in the U.S. are food use and beer.

During the last decade, U.S. rice used for direct-food and processed-food was about 80 percent while rice used to brew beer averaged 10 percent, the remaining 10 percent was used for livestock feeds.[3]

Rice is, however, a major crop for the U.S. in international trade; it accounted for about 30 percent of all rice in the world trade in 1978.[4] Therefore, exports have been the most important outlet for U.S. rice industry.

During the 1970's, U.S. rice exports accounted for about 60 percent of total production. The two types of U.S. rice exports were commercial rice shipments and shipments under U.S. government aid programs. Most commercial sales of rice went through normal business channels, which ranged from 60 percent over the 1970's to about 90 percent of total rice exports in the 1980's.[3] The U.S. government also provided food aid programs to support rice exports. The most important program by far is the PL-480 program which assists in the long-range improvement of economic systems in developing and less-developed countries by providing donations, aid and long-term credit.

After World War II, rice production in the U.S. almost doubled the level of the prewar period.[5] Excess rice production exerted serious downward pressure on rice prices since 1950's. Beginning with the 1974 crop, the rice acreage control programs were lifted and, as a result, U.S. rice production has increased from 1.9 million acres for the 1960-73 period to 2.8 million acres for the 1974-80 period.[1] The U.S. rice ending stocks rapidly increased from an average of 9.6 million hundredweights for the 1961-74 period to 32.4 million hundredweights for the 1975-80 period.[1] The U.S. rough rice production has increased from 3.8 million metric tons to 8.3 million metric tons in 81/82 crop year, a 118 percent increase. The production was around 5-8 million metric tons in 1980's and 1990's.[2]

As rice production increased rapidly both in the U.S. and in foreign countries (table 1), U.S. rice industry has been facing a question of how to keep an appropriate share in the world market since exports were the most important outlet for U.S. rice industry. The expanded rice supplies in the world market caused downward pressure on export prices (table 2). The U.S. Department of Agriculture has responded by trying to use various policies to stabilize prices and farm incomes in terms of loans and price supports. With domestic farm programs supporting prices above the world level, U.S. rice exports fell about 26.7 percent in the early 1980's and from 3 million metric tons in 1980 to 0.1 million metric tons in 1989 and 1990,[2] a 95 percent decrease, while some other countries' exports have rapidly

Overview

increased in the same periods (table 3). Since changes in transportation costs are very important in determining the routes over which rice should be shipped, and the export routes that are determined by shipping costs affect the routing of rice from the U.S. to importing countries, ocean transportation costs have an indirect effect on domestic routing of shipments and consequently on the geographic price surface for rice in the United States. Therefore, transportation cost for shipping rice from the U.S. to world market has been one of the factors affecting the competitive position of U.S. rice producers in the world rice market, and hence affecting the magnitude of U.S. rice exports. It has great influence on the U.S. grain economy as a whole.

Table 1 Trend of Rice Production in Major Rice Exporting Countries *(Marketing Year)*

Year	USA	Thailand	China	Pakistan	Burma	World
			(million metric tons)			
70/71	3.8	12.1	97.5	20.0	8.2	290.0
75/76	5.1	14.5	120.0	3.5	8.6	352.3
80/81	6.6	17.4	139.9	4.7	13.3	397.4
81/82	8.3	17.8	144.0	5.1	14.1	412.4
82/83	7.0	16.9	161.2	5.2	14.4	419.8
83/84	4.5	19.5	168.9	5.0	14.4	452.7
84/85	6.3	19.9	178.3	5.0	14.8	469.0
85/86	6.2	19.7	168.5	4.4	14.9	465.5
86/87	6.1	18.0	171.1	5.2	14.8	466.8
87/88	5.9	18.0	169.1	4.6	12.5	484.0
88/89	7.3	21.3	169.1	4.8	12.5	488.8
89/90	5.1[a]	13.3	126.1	3.2	8.1	344.1
90/91	5.1	11.3	132.5	3.3	8.2	352.3
91/92	5.0	13.5	128.7	3.2	7.4	349.5
92/93	5.7	13.1	130.4	3.1	7.8	352.5
93/94	5.0	12.2	124.4	4.0	8.8	350.4

Sources: USDA, *Agricultural Statistics,* 1970-94; USDA, *Foreign Agricultural Circular-Grains,* various issues.
[a]Rough rice was used until marketing year 1988/89, after which milled rice was used.

Table 2 Trend of Rice Export Price Index (1957-59=100)

	1978-80	1981	1982	1983
	(dollars per ton)			
Total index	280	345	242	216
Private trade	278	332	231	226
Bilateral contracts	282	360	255	204
Long/medium grain	273	334	225	206
Round grain	319	400	327	267

Source: FAO, *Commodity Review and Outlook*, 1983-84. FAO Economic and Social Development Series, No.29, Rome, 1984, p. 50.

Table 3 Trend of Rice Exports in Major Rice Exporting Countries (on milled basis)

Year	USA	Thailand	China	Pakistan	Australia	World
			(million metric tons)			
1970	1.7	1.1	0.9	0.9	0.1	7.0
1975	2.1	0.9	1.4	0.5	0.2	7.2
1980	3.0	2.7	1.0	1.0	0.3	12.5
1981	3.0	3.1	0.6	1.1	0.4	13.0
1982	2.5	3.6	0.5	0.8	0.5	11.6
1983	2.3	3.7	0.6	1.3	0.3	11.9
1984	2.1	4.5	1.2	1.1	0.4	12.6
1985	1.9	4.0	1.0	1.0	0.4	11.5
1986	2.4	4.3	1.0	1.2	0.4	12.8
1987	2.4	4.4	1.0	1.2	0.4	12.6
1988	2.2	4.8	0.7	1.0	0.4	11.1
1989	0.1	6.0	1.2	0.0	0.0	14.2
1990	0.1	3.9	0.3	0.0	0.0	11.7
1991	2.2	4.0	0.7	1.3	0.5	12.1
1992	2.1	4.8	0.9	1.4	0.5	14.1

Sources: USDA, *Agricultural Statistics*, 1970-93; USDA, *Foreign Agricultural Circular-Grains*, various issues.

REVIEW OF LITERATURE

Empirical studies on transportation costs in international trade are very limited due to the lack of available data. However, since the growing importance of the transportation costs to the world economy—they influence the directions and magnitudes of trade flows and gains—is increasingly realized, economists have recently begun analyzing the effect of transportation costs on international trade and incorporating them into trade models.

Some researchers have studied the costs of transportation for a specific commodity. Others have analyzed the optimum transportation patterns for moving certain commodities between locations within the country. For instance, a transshipment model of the U.S. rice industry was first developed by Holder et al.[6] in 1971 to determine the optimum location of rice mills, flow of rough rice from the representative dryer locations to each optimum mill location, and flows of milled rice from the optimum locations to the specified demand points.

U.S.D.A.[7] published an article analyzing the effects of cost-of-service transportation rates on the U.S. grain marketing system. This research can be briefly outlined as follows: the costs of transportation services to carriers (truck, rail, and barge) for transporting grain and flour were developed by synthesis for specified U.S. regions from secondary data. These costs served as input data for an operational mathematical transshipment model used to analyze effects of a cost-of-service transportation rate structure on the U.S. grain marketing system.

Stennis and Pinar[8] studied the economic effects of transportation costs on the flows of international cotton shipments. One of the results revealed that a cost-insurance-freight (C.I.F) valuation base placed a disproportionate burden on countries which had relatively higher transportation costs because of geographically-disadvantaged locations relative to other countries. This, in turn, would affect the net gains from international cotton trade.

A transportation study on short-run cost functions for Class II railroads was developed by Cheaney, Sidhu and Due[9] which used an econometrics model to determine the short-run responsiveness of various cost categories to volume changes. Specifically, the study was concerned with the cost functions of a sample of ten Class II railroads for a period of ten years. The basic conclusion was that, in the short turn, additional traffic on light traffic lines would significantly lower train operating and maintenance-of-way costs, and therefore improve the financial viability of the lines.

Moser and Woolverton[10] developed an estimate of barge transportation costs for grain and fertilizer. Their study provided useful information on the costs of barging grain on the Mississippi and Missouri Rivers. Legislation affecting cost of service was reviewed. Rate setting procedures used by the barge industry were examined. The report included a discussion of the methodologies various researchers had used in attempting to synthesize barge cost-of-service, and a presentation of the results.

Sharp and McDonald[11] conducted a study on the impact of vessel size on an optimal system of U.S. grain export facilities. This study determined the impact of ocean vessel size on the transportation cost of U.S. exports of heavy grain to seven foreign demand regions and the associated impact of vessel size upon the U.S. export grain facility requirements. It concluded that such a system must incorporate the utilization of large-scale, low-per-unit-cost vessels which would enable the U.S. to maintain a competitive position in the world trade of heavy grains by minimizing transfer costs.

U.S.D.A. Agricultural Marketing Service[12] conducted a study to determine the direct labor, equipment, and material costs to harvest and deliver watermelons from growing areas to retail stores by the bulk and bin methods that were used. Two systems of handling, bulk and bin, were commonly used in marketing watermelons. The bin system was found to be the least expensive method of handling. This system began after the watermelons were moved from the field by trucks. The melons were loaded onto the over-the-road trailer for shipment to the wholesaler. This system, though least expensive, was not viable because produce managers could not usually use all of the bins received in the process.

Davis[13] developed a transportation model to determine a least-cost shipping pattern for U.S. grain exports. In the model, the author used the data developed on the cost per ton of shipping grain for three bulk grain vessel sizes from U.S. ports of origin to specific ports of destination. In addition to estimating the cost per ton of grain shipments, this study analyzed heavy grain (wheat, corn, soybeans, sorghum grains) shipments for the years 1958 and 1966, and determined the types of costs involved in shipping grains. The resulting transportation model indicated that the Gulf Region was the principal exporting region in the U.S., while the other three regions exported to a few selected market areas. Another finding was that if 50 percent of government sponsored shipments did not have to be carried on

Overview

U.S. flag vessels, over $200 million dollars per year could be saved in transportation costs.

Pinar[14] analyzed the effects of ocean transportation costs and tariff barriers on the flows of international cotton shipments. A transportation model, within a linear programming framework, was utilized to obtain the optimal flows in the international cotton market. A comparison of the actual cotton flows and optimal cotton flows was made to determine if additional net social gain could be realized with the optimal flow of cotton in world trade.

Olechowski and Yeats[15] demonstrated that more than one-third of the products had freight factors (transportation cost divided by f.o.b. price) of over 20 percent which emphasized the potential importance of transportation costs as a trade barrier for international trade.

Binkley and Harrer[16] conducted an econometrics analysis to examine the major factors that determine the ocean freight rates for grain exports. They indicated that a country's competitive position in the world grain market depends upon its competitive advantage in shipping and in production. One of the most important implications of the study was the relationship between international shipping and the country's comparative position in the world grain market. They suggested that the effects of changes in international transportation costs on trade is itself worthy of study.

U.S. SHIPPING INDUSTRY

Within an international context, the need for transportation services to move goods from point of surplus to point of deficit demand requires the existence of a substantial shipping industry. The structure of this industry is quite complex. Ships carry goods of widely differing values and handling characteristics to all ports of the world. This in turn requires a variety of ships, terminal facilities, charter arrangements and other services to facilitate the smooth transport of these goods.

There have been around 150 countries, not including their dependencies and colonies, with which the U.S. has carried on its foreign trade. Except for Canada and Mexico, ships and airplanes have been the only means of transportation. And even in trade with Canada and Mexico, ships carried large quantities of cargoes in both directions.

The U.S. shipping industry includes basically two separate business categories: liner and bulk. Liner sectors operate mainly containerships and other general cargo ships in a regular scheduled service carrying various cargoes from port to port at set rates. The bulk-shipping companies deal with large tonnages of single commodities by operating or owning a fleet of ships for one cargo. Bulk business includes petroleum corporations that operate tanker fleets, and independent bulk ship operators that operate tankers, dry-bulk carriers (ore, coal, grain) and combination ships. The U.S. shipping industry is also divided by flags of registry. The U.S. merchant fleet is considered to include U.S.-flag privately owned, self-propelled vessels of over 1,000 gross tons. This definition includes basically all U.S.-flag ships involved in international trade as well as in the domestic coastal and offshore trades.[17]

The U.S. shipping industry can be extended to include the fleets owned by U.S. corporations but registered in other countries, which consist basically of tankers and dry-bulk carriers. About 36 percent of the Liberia-flag fleet and 17 percent of the Panamanian-flag fleet are owned by U.S. shipping companies.[17,27]

The U.S.-flag fleet size has been largely influenced in the past half century mainly by the massive ship building programs associated with World War I and World War II. The U.S. fleet's tonnage has changed greatly in the last 50 years.

The United States pioneered containerization, a system which has enabled steamship companies to reduce port time and cargo handling costs. United States Lines *AMERICAN ASTRONAUT* in the U.S. container fleet, is the largest and most modern in the world, comprising more than 25 percent of the world container fleet. The U.S. shipping industry also consists of two separate competitive fleets: the subsidized fleet and non-subsidized fleet. The non-subsidized fleet receives no direct subsidy but may receive indirect subsidy such as unusually high freight rates. The subsidized fleet sector operates the best ships, enjoys the best earnings, receives funds and are protected from competition.

Table 4 lists U.S.-flag merchant fleet by type of ships. The U.S. liner fleet is the predominant sector in international trade. Most liner shipping firms belong to steamship conferences, which set rates and establish rules and operating agreements for their members. The U.S.-flag tanker and dry bulk fleets dominate the U.S. domestic trade. The three largest ship-operating firms own more than 50 percent of the total tonnage. Seven of the major ship firms operate their international-trade-engaged vessels under the U.S. Maritime Administration's Operating Differential

Overview

Subsidy Program, and up to 20 percent of each firm's cost deferential with foreign-flag vessels shipping on the same trade route is covered by direct subsidies.[18] Only Sea-Land, one of the largest firms, does not receive direct subsidies.

Table 4 U.S.-flag Privately Owned Merchant Fleet
         (Oceangoing Ships, 1,000 Gross Tons and Over)

	Number of Ships	Deadweight tons
General cargo	240	4,312,153
Breakbulk/partial Container	104	1,404,688
Containership	97	1,868,274
RO/RO-vehicle carrier	18	274,043
Barge carrier	21	765,148
Bulk cargo	18	618,019
Tankers	233	14,220,469
Special product/liquefied natural gas (LNG)	33	1,601,551
Other (coastal, passenger)	17	110,396
Total	781	25,174,741

Source: "Ship Register," Military Sealift Command, Department of the Navy, Washington, D.C., 1983. Cited in "An Assessment of Maritime Trade and Technology." Congress of the United States, Office of Technology Assessment, 1983.

U.S. liner business in international trade has increased about 30 percent in tonnage over the last 10 years, but the U.S.-flag industry has remained rather constant in tonnage. During the first half of the 1980's the U.S. liner industry shrank substantially due to the worldwide recession and decreased overall shipping volume in major trades. Some smaller companies were facing very difficult financial positions while a few of the larger operators were expanding shipping services and building new containerships to modernize their fleet.[17]

The U.S.-flag tanker fleet in foreign trade is small and attracts little business in the severely over-tonnage international markets. The U.S. dry-bulk fleet engaged in international trade consists of 23 vessels.[17] Most of the ships continue operating because they carry Government-preference cargoes, principally AID shipments, where they do not have to compete with foreign ships (table 5). A very important characteristic of the U.S. dry-bulk trades is the intense price competition.

Table 5 U.S. Dry-Bulk Carriers in International Trade

Name of vessel	Type	Deadweight tons	Year Built
Jade Phoenix	BO (Bulk/Oil)	63,200	1982
Marine Princess	Bulk	52,565	1967
Overseas Harriette	Bulk	25,541	1978
Overseas Marilyn	Bulk	25,541	1978
Point Manatee	Bulk	15,316	1944
*Pride of Texas	Bulk	35,389	1981
Seadrift	Bulk/Oil	15,155	1942
*Spirit of Texas[a]	Bulk	32,100	1982
*Star of Texas[a]	Bulk	36,614	1982
Sugar Islander	Bulk	29,648	1973
Tamara Guilden	Bulk	23,800	1961
*Ultramar[a]	OBO(Ore/Bulk/Oil)	82,199	1973
*Ultrasea[a]	OBO(Ore/Bulk/Oil)	82,199	1974
Betty Wood	Bulk (Tug/Barge)	23,751	1973
Calrice Transport	Bulk (Tug/Barge)	25,000	1976
Jamie A.Baxter	Bulk (Tug/Barge)	24,372	1977
Moko Pahu	Bulk (Tug/Barge)	25,931	1982
Total (operating)		764,051	
*Golden Phoenix[b]	BO (Bulk/Oil)	129,000	1983
Ogden Parana[c]	Bulk	45,000	1983
Ogden Trent[c]	Bulk	45,000	1983

Sources: U.S. Department of Transportation, Maritime Administration, Office of Trade Studies and Subsidy Contracts, Division of Statistics, March 8, 1983. *"An Assessment of Maritime Trade and Technology."* Congress of the United States, Office of Technology Assessment, 1983.
*Vessel built with CDS.
[a]Currently operating in preference trades under Sec. 614 of Merchant Marine Act.
[b]Under reconstruction. Former LNG carriers.
[c]Under construction.

If U.S. dry-bulk export trade follows the Wharton Econometrics forecast of over four percent per year between 1980 and 2000, the U.S. will require additional fleet service in trade.[17] Table 6 shows trade-growth estimation for policy changes, the future fleet growth will depend heavily on the business strategies of large bulk shippers.

Table 6   U.S. Dry-Bulk Export and Import Trades

Commodity	1980	1990
	(million tons)	
Coal	72.8	133.0
Iron ore	25.5	57.7
Grain	97.2	143.6
Alumina/bauxite	20.2	22.6
Phosphate rock	14.8	19.6
Rice	2.0	2.6
Sugar	4.8	5.1
Sorghum	5.4	6.3
Soybeans/Meal	28.8	30.0
Forest products	27.6	35.7
Fertilizers	13.4	17.8
Potash	1.2	0.2
Sulfur	1.7	0.8
Chrome ore	1.2	0.9
Gypsum	6.8	9.5
Manganese ore	1.2	0.9
Iron/steel scrap	10.7	11.4
Petroleum coke	8.0	13.0
Other	4.9	4.8

Source: U.S. Maritime Administration, FACS, 1982. Cited in *"An Assessment of Maritime Trade and Technology."* Congress of the United States, Office of Technology Assessment, 1983.

FLAG OF REGISTRY

International competition in merchant shipping has been complicated by flag of registry. All vessels are registered in a nation and are owned by an individual or a company incorporated in the nation of registry. All vessels are under the jurisdiction of the maritime authority of the nation of registry and are bound by its laws and regulations. Since the degree of control exercised by sovereign nations varies greatly, there is a distinction made between a vessel registered under the national flag or under a flag of convenience. All shipping firms operating under a given registry face similar cost structures. Cost inequality among countries is solely the product of their respective maritime policies which apply equally to all companies of a given flag.[19]

The most common policies associated with a flag of registry are policies regarding where ships can be purchased, who may work on these ships, and how these ships are taxed and regulated. Most countries involved in international sea transport apply similar policies for the first two. Ships are bought where the price is most favorable. Crews are hired where it appears most advantageous. However, differences exist among countries in respect to taxation and regulation. Some countries, known as "convenience" countries, allow easy registration with minimum taxes and regulations. These countries are also known as "open registry" countries because they allow the shipowner from other countries to register vessels under their flags to take advantage of the absence of taxation and regulation. The shipowners have the flexibility to use crews of any nationality, to contract the ships in any country, and to operate outside the framework of their own national laws and regulations.[19]

Flags of convenience emerged around 1939 when the United States was following a neutrality policy but wished to help its allies with shipments of supplies. The major countries that currently permit open registries are Liberia, Panama, Cyprus, Singapore, and Somalia. They offer open registries in order to increase employment of their citizens in the shipping industry and gain a larger share of the cargo market.

Open registry has been most attractive to the U.S. shipowner because U.S. maritime policies prevent the U.S. shipping industry from being competitive in international shipping. The term often given the U.S. flag-of-convenience fleet is the "U.S. effective-control" fleet, because it is U.S.-owned, it is effectively under U.S. control and can be considered as part

Overview

of the U.S. fleet. Table 7 lists the ships that were owned by the U.S. companies and registered under foreign flag in 1982. The total tonnage is about twice the U.S.-flag fleet.

Table 7 The U.S. Effective-Control Fleet

	Number	Total (000 Deadweight-ton)	Average
Total	466	47,221.8	101.33
General cargo	73	525.7	7.20
Breakbulk/reefer	52	334.4	6.43
Containership	10	25.5	2.55
RO/RO	6	35.0	5.83
Barge carriers	5	130.8	26.16
Bulk	106	6,466.6	61.01
General bulk	76	3,537.9	46.55
Combination, ore/bulk/oil	30	2,928.7	97.62
Tanker	259	39,426.7	152.23
Special product/LNG	27	793.3	29.38
Passenger	1	9.9	9.90

Source: Federation of American Controlled Shipping, March 1983. Cited in "*An Assessment of Maritime Trade and Technology.*" Congress of the United States, Office of Technology Assessment, 1983.

U.S.-flag ship costs are substantially higher than foreign-flag costs for both ship acquisition and operation. Therefore, U.S. operators have had difficulty competing in the world market. Data from *Man Ad* for 1982[17] indicate that new construction costs for containerships are two to 2 1/2 times higher in U.S. shipyards than in comparable foreign yards, such as Japan. While all operating expense categories are higher for the U.S. shipping operator than the foreign operator, crew costs are a major item of difference, particularly due to different manning scales rather than per-man wages. Foreign crew costs for containerships range from one-half to one-sixth of equivalent U.S.-flag crew costs (figure 1). Most U.S.-flag crew sizes exceed

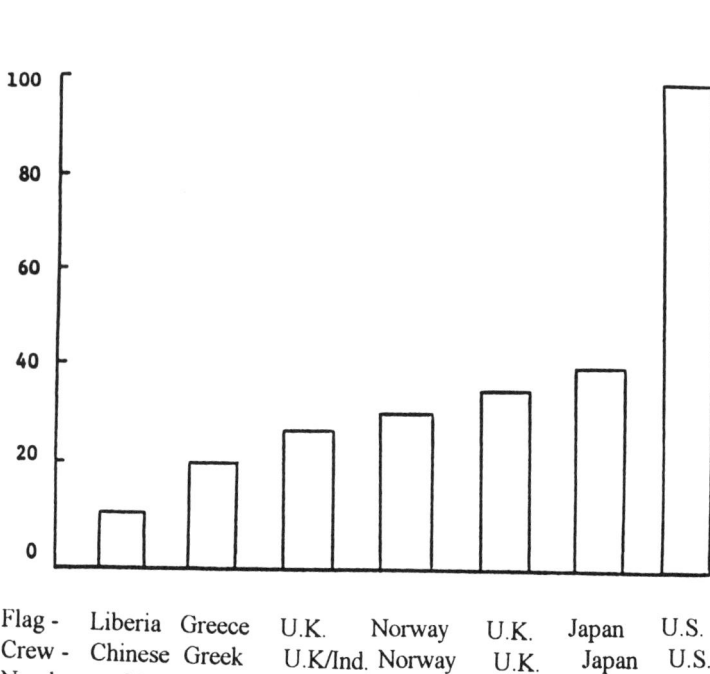

Figure 1. Relative Costs for Crew for a Typical Containership Operating Under Various Flags

Source: "Comparative Operating Costs for U.S. and Foreign-Flag Ships." by Paul Ackerman, Presented at SNAME Ship Cost and Energy Symposium. October 1982. "An Assessment of Maritime Trade and Technology." Congress of the United States, Office of Technology Assessment. 1983.

Overview

foreign-flag crew size, which is much more important than the wage rates. U.S. Department of Transportation (DOT) data show a ratio of U.S. daily costs v. comparable average costs for Organization for Economic Cooperation and Development (OECD) countries for a 26-man dry-bulk ship crew of about three to one.[17]

The U.S.'s high standard of living mainly accounts for the higher wage rates. However, European and Japanese crew wages have also been increasing substantially, and they are also facing a competitive disadvantage compared with LDC crews. Many European and Japanese ship operators have therefore countered the wage rate problem with smaller crew size, hiring foreign crews and using more automation—an approach only now beginning to be used by U.S. shipping operators.[17]

Table 8 Monthly Average Compensation* for Selected Ship Workers by Country.

	Master	Second Engineer	Seaman
	(dollars)		
United States	17,387	8,212	3,301
Japan	9,705	4,820	3,643
West Germany	7,401	4,174	2,200
Sweden	8,695	4,813	2,605
Denmark	5,945	2,899	2,428
Korea	2,800	905	644
Taiwan Chinese (Panamanian flag)	2,505	1,295	770
Hong Kong (Liberian flag)	2,708	1,293	721
Ghana	2,062	1,610	442

Source: Morgan Guaranty Trust Company. Cited in *"Determinants of Ocean Freight Rates for Grain,"* by Christen Ruth Olson. Unpublished M.S. thesis, North Carolina State University, 1983.
*These figures include pay and fringe benefits.

Subsistence, stores and supplies are proportional to crew costs. U.S. ship maintenance and repair costs are also higher than foreign competitors (table 8). It usually costs more to repair a ship in a U.S. yard, and a U.S. operator who repairs in a foreign yard must now pay a 50-percent ad valorem tax. U.S. insurance costs are also higher, reflecting the higher capital costs of U.S. ships. Another main factor for high U.S. shipping cost is fuel. Expenses for crew and fuel account for a significantly higher proportion of overall operating costs for bulk ships than for liners, limiting the opportunities to reduce the cost differential through efficiency improvements in other operating components. Therefore, it is necessary to increase ship energy and efficiency and reduce crew costs which justifies modernizing the U.S. fleet. When these two goals are achieved, the U.S.-flag ships will become more competitive with the rest of the world.[17]

If a more moderate growth rate occurs in future trade, U.S. shipping operators will be forced to compete with rapidly growing foreign-flag fleets for the limited cargo shipment available and will need to continue increasing service efficiency and capability.

THE U.S. CARGO PREFERENCE POLICIES

Generally speaking, because of the restrictions and regulations under U.S. maritime policies which have resulted in higher costs, the U.S. maritime fleet is usually not internationally competitive. However, some of the U.S. vessels —even with their higher capital cost, higher wage cost and higher fuel cost —are able to compete in some trades with their foreign-flag counterparts. The U.S. government subsidization is one of the main reasons for the U.S. fleet to continue to operate. In the past, U.S. construction subsidies were available to mitigate higher construction costs. Operating subsidies also helped. In addition, U.S. preference cargoes are available to cover some added costs.[17,20]

Many major U.S. liner trades are more imported than exported. U.S. flag liner operators have historically reduced the economic impact of this imbalance by carrying government cargoes outbound. The U.S. government had been applying a policy of granting preference to U.S. carriers on its own cargoes. The major components are Agency for International Development (AID) and Import-Export Bank cargoes and military cargoes.[17]

Table 9 shows the importance of preference cargoes to U.S.-flag liner operators in 1982. The impact of these cargoes on individual carriers varies widely. In some cases up to one-third or more of total revenue is derived from such carriage. The preference cargoes are significant enough to a U.S.-flag operator that the resulting higher utilization of his ships will bring down unit costs by a sizable amount.

Table 9 Importance to Liner Vessels of Carriage of Government Preference Cargoes

	1978		1979		1980	
	million $	%	million $	%	million $	%
Total U.S. operator revenue	3,105	–	3,707	–	4,308	–
Preference revenue- civilian cargo	283	9	266	7	294	7
Preference revenue- military cargo	201	7	376	10	401	9
Total preference revenue	484	16	642	17	695	16

Source: U.S. Maritime Administration, Office of Policy and Plans, November 1982. Cited in *"An Assessment of Maritime Trade and Technology."* Congress of the United States, Office of Technology Assessment, 1983.

Most experts agree that some U.S.-flag liners can be productive and competitive in the world market despite some cost disadvantages. Government policy can mitigate important cost disadvantages without direct subsidy. On the other hand, a number of subsidized U.S.-flag operators depend heavily on direct subsidy payments for their financial survival. These operators require productivity improvements and substantial future cost reductions to meet foreign competition. If future subsidies are reduced or eliminated, attention to successful productivity improvements for these companies must receive high priority.[17]

Government subsidization is important because U.S. maritime laws contain restrictions which differ from most countries engaged in international sea transport. The U.S. shipowners are, with a few exceptions, required to purchase their capital equipment within the U.S. This has raised costs substantially. Also, U.S. shipowners may employ only U.S. citizens as crew.

The wages of U.S. seamen are by far the highest in the world.[17] Because of those restrictions, U.S. ship operators are forced to either register their ships in open registry countries or become heavily dependent on government subsidies and other forms of protection for survival.

In addition to direct subsidies which apply mainly to liner shipping operations, other measures of protection are provided by the U.S. government. Through cabotage laws, foreign flag ships are prohibited from carrying domestic cargoes. Through cargo preference laws, certain cargoes are mandated to move on U.S.-flag ships.

The practice of restricting certain cargoes to U.S.-flag ships began with the 1904 law requiring that all military cargoes be moved on U.S.-flag ships. In 1948, Congress passed the first cargo preference provision for aid cargoes. This practice continued on an annual basis until 1954, when Public Law 664 made it permanent. This law, Section 901 of the Merchant Marine Act, required that 50 percent of all U.S. government-impelled cargoes be shipped on U.S.-flag vessels.[19]

The U.S. 1985 Food Security Act tremendously changed cargo preferences laws which formerly required that at least 50 percent of government-sponsored shipments be made on U.S.-flag ships. This Act clarifies that requirement and mandates a gradual increase in the share of particular exports, mostly feed aid, that must be shipped on U.S.-flag vessels. The cargo preference requirements do not apply to certain specific commercial agricultural export programs such as export credit, credit guarantee, etc. In 1986 and 1987, it required that 60 percent and 70 percent food aid exports be shipped on U.S.-flag vessels respectively. And in 1988 and thereafter, at least 75 percent of such exports must be shipped on U.S.-flag vessels.[21]

The AID and U.S.D.A. (U.S. Department of Agriculture) are by far the largest shippers of nonmilitary government transported cargo, of which agricultural commodities make up a large part. Cargo preference laws served as a type of quota in that they restricted foreign competition in certain markets, reduced the supply of shipping services, and thus maintained rates at levels high enough to allow U.S.-flag operators to stay in business. The importance of cargo preference to the U.S. maritime industry is significant. In 1980, revenue from the carriage of preference cargoes totaled $1.1 billion for all U.S. operators. Liner operators received 16 percent of all revenues under the programs.[17] The continued existence of cargo preference laws has usually been justified by its proponents on the basis of three considerations: infant industry, balance of payments and national defense.

Overview

As previously mentioned, there are large difference in unit transportation cost between U.S.-flag vessels and foreign-flag vessels. The agricultural cargoes shipped in U.S.-flag vessels under the provisions of the cargo preference act operate at substantially higher rates than do cargoes on foreign-flag ships. U.S.-flag ship operators are not competitive with other major shipping nations (nor are they competitive with the foreign-flag U.S.-controlled fleet). Cargo-preference policies have been expanded to support the commercial U.S. fleet trades. However, it does not appear that such incentives would provide sufficient support if they were reinstated at the same level today.

RICE TRANSPORTATION VESSELS

Rice is exported on three general types of ships: cargo liners, tanker vessels and tramp steamers. Table 10 shows the share of shipments of U.S. rice exports by these three types of vessels from 1981 to 1986.

Liners

Cargo liners are ships traveling a fixed route in accordance with a predetermined time-table, in a regular scheduled service carrying various cargoes from port to port at set rates, much like a railroad or trucking operation. Liner owners usually sell space on a vessel by the freight-ton to a number of different shippers at predetermined rates. Two types of rate schedules are used by liners: class tariffs and commodity tariffs. Under a class tariff, products are carried at a rate determined for that class. Under a commodity tariff, each good carried is given a separate rate. When two or more liner companies serve the same route, competition between them is invariably limited by arrangements covering freight rates to be charged and other aspects of the competitive relation.[22] This type of arrangement among liner owners is known as a liner conference.

Table 10   Share of Shipment of U.S. Southern Region Rice Exports by Different Vessels

Year	Liner tons	Liner %	Tanker tons	Tanker %	Tramp tons	Tramp %
1981	583,341	25.0	69,850	3.0	1,699,975	72.0
1982	647,600	31.5	78,735	4.0	1,329,081	64.5
1983	588,586	35.8	118	0.1	1,052,572	64.1
1984	679,515	57.5	0	0.0	502,462	42.5
1985	626,856	48.8	0	0.0	657,694	51.2
1986	491,381	22.2	2,922	0.1	1,720,110	77.7

Source: U.S. Department of Commerce, Bureau of the Census, Unpublished Foreign Trade Statistics, 1981-1986. Bureau of the Census, U.S. Department of Commerce, Washington, D.C.

The largest and most prominent liner companies are increasingly engaged in cargo transportation between inland locations in which ships serve only as links in an overall transport system. The rates published by cargo liners for large quantities of bulk commodities are termed open rates and are determined by negotiation between the ship owners and prospective shippers. These carriers usually offer a scheduled service; the bulk commodities which they carry are known as liner parcels.[22]

Shipments of rice on liners have been significant in past years. During the 1980's, cargo liners accounted for 22.2 percent to 45 percent of U.S. rice exports.[23] The U.S. liner fleet has maintained a relatively large share of U.S. rice export trade despite effective foreign-flag competition. This is partly because of successful productivity improvements by major operators and also because of Federal subsidies that helped to maintain U.S. liner fleet's cargo share position.[17]

Tankers

Tanker operators usually handle large tonnages of single commodities by operating one or a fleet of ships especially designed for one cargo. Tankers can therefore take advantage of economies of size. However, the advantages of tankers are minimized and may even be entirely offset by excessive turn-around times in loading and discharging. Most ports importing rice have an unloading capacity insufficient to take advantage of tankers. Ports must also be able to offer safe deep water anchorage for the big ships. In the shipment of U.S. rice export, tankers are the least important vessels used among all the three types of ships. The largest amount of rice export carried by tankers in the 1980's was four percent.[22] The U.S.-flag foreign trade tanker fleet is small and is attracting little business in the severely over-tonnage international markets. Due to the lack of opportunities in the world market, much of the U.S. subsidized fleet took advantage of a provision allowing tankers to enter the domestic trade.[19]

Tankers, as big bulk carriers, will become important in the shipment of rice exports when ports can offer high discharge rates and safe deep water anchorage. They also depend largely on the business strategies of large bulk shippers and the ship owners and operators who carry the cargo.[19]

Tramps

Tramps or tramp steamers, are ocean carriers offering irregular service with no fixed ports-of-call. They travel no predetermined route. Bulk agricultural commodities, such as grain and fertilizers, are their chief cargoes. Tramps' rates are determined by negotiations between the shipper and carrier, with a shipbroker usually serving as a intermediary. The agreement that stems from such negotiations is referred to as a charter party. Tramp owners charter their vessel to shippers either on a voyage basis, in which case the contract is usually for one voyage and for a particular good, or on a time basis where the contract is for a specific period of time. Single voyage charters are generally referred to as open market fixtures while time charters are referred to as negotiated fixtures.[28]

Tramps by far are one of the most significant transportation means for carrying rice exports from the U.S. to international markets. Tramp vessels accounted for 72 percent of rice exports from U.S. southern ports in 1981 and 77.7 percent during 1986. Rice chartering on tramp ships averaged 1.22 million tons a year from 1981 to 1986.[23]

CARGO VESSEL CHARTERS

A charter is a contracted arrangement based on the mutual commercial interests of a charterer, who requires a vessel to meet his transportation needs, and an owner who places his vessel at the disposal of the charterer. The practical system of chartering a ship is not very different from any other major hiring or subcontracting operation except that the business is very international. The "shipowner" comes to the market with a ship available, free of cargo. The ship will have a particular speed, cargo capacity, dimensions and cargo handling gear. The vessel will probably have existing contractual commitments, which will determine the date and location at which it will become available. The "shipper" or "charterer" hires a volume of cargo to transport from one location to another. The quantity, timing and physical characteristics of the cargo will determine the type of shipping contract he requires. The "brokers" and "agents" specialize in setting up deals between shipowner and charterers, though some shipowner may pursue contracts directly, i.e., they approach a shipper and arrange a charter themselves.[25]

There are four broad categories of charters in general use. They are voyage charters, contracts of affreightment, time charters and bareboat (demise) charters. From the shipowner's point of view, the main difference between these types of charters is the degree of owner involvement in the shipping operation and the division of costs.

Voyage Charter

A single-voyage charter is the most common contractual arrangement. As its name implies, it is a contract for the carriage of cargo not for a period of time but at a stipulated rate per ton, on one voyage between two ports on a fixed range of ports. The owner receives a freight payment for the movement of a cargo between ports, from which he must pay all voyage and operating costs. The charterer assumes no responsibility whatever for the navigation of the vessel or the custody or safety of the cargo.[25]

Overview

Contract of Affreightment

Sometimes the shipowner may wish to do business on the basis of a fixed price per ton of cargo transported, but may not want to tie himself contractually to using a specific vessel. If he operates a fleet of ships, he may want to switch cargo between different vessels in order to gain the best possible operating pattern and lower charter rate. Similarly, the shipper sometimes may have several regular cargoes to ship and may wish to arrange for the shipment in one contract and leave the details of each voyage to the shipowner. In these cases, a contract of affreightment provides the shipper with "no hassle" transport for his cargo, while the shipowner has flexibility to deploy his fleet.[25]

Time Charter

A time charter is a contract for the use of the cargo-carrying space in a vessel over a specified period of time, which could be anytime from that taken to complete a single voyage or for a period of months or years. The ship owner bears all the costs and risks of vessel operation and the charterer is responsible for all cargo handling charges and voyage costs (i.e. bunkers, port charges and canal dues). With a time charter, the shipowner continues to pay the operating costs of the vessel (i.e. the crew wages, maintenance and repairs) during the whole period and must provide a warranty regarding the vessel's performance in terms of speed and fuel consumption. The charterer has a clear basis for preparing the ship budget as he knows the ship operating costs from experience and is in receipt of a fixed daily or monthly charter rate.[25]

Bareboat Charter (Demise)

In a bareboat charter, also called a time charter by demise, the owner delivers the vessel bare to the charterer who carries out all the functions of vessel operator normally performed by the owner. Under this arrangement, the shipowner purchases the vessel and hands it over to the charterer who manages the vessel and pays all operating and voyage costs. Once the charterer accepts the vessel and starts to pay the monthly payments to the owner, the stream of payments continues without interruption for the duration of the charter period and the owner is not active in the operation of the vessel and does not require any specific maritime skills.[25]

TERMS OF SHIPMENTS

In international trade markets, there generally exist two categories of terms of shipment. One category refers to the contractual arrangement between the exporter and importer of the goods as to who is responsible for bearing the risks and costs associated with the transport of the goods from the exporter's point into the importer's point. This category of terms of shipment includes the following terms:[26]

(1) *C.I.F. (cost, insurance, freight)* Under C.I.F. contractual terms, the seller (exporter) is responsible for contracting shipping services, procuring insurance and providing the buyer with certain official documents. That is to say, the exporter is responsible for the goods traded until they are unloaded at the importer's destination port.

(2) *F.O.B. (free-on-board)* Under F.O.B. contractual arrangements, the buyer (importer) is required to contract shipping services and procuring insurance. The seller (exporter) is responsible only for delivery of the goods on board a specified vessel at a time agreed upon.

(3) *C.&F. (cost-and-freight on importer dock)* This is a variation of C.I.F. contractual terms.

(4) *F.A.S.(free-along-side bags on warf)* This is a variation of F.O.B. contractual arrangements.

Recently, because of increased cost and largely government-involved actions in trade, F.O.B. contractual arrangements have been increasingly preferred.

In the procedure of chartering ocean freight, shipowner and charterers must decide who is to pay for the loading and unloading of the ship's cargo. This is specified in a ship charter as the terms of shipment of another category. The responsibilities for loading and unloading a ship's cargo are specified under three types of shipment terms employed in this category:[26]

(1) *Free-in-and-out* Under this term, the charterer is responsible for both loading and unloading of the cargo.

(2) *Free-discharge* When this term is specified in a ship charter, the charterer is to pay for the discharge costs, and the shipowner pays for the loading.

(3) *Gross terms* This shipment term requires the shipowner to pay both loading and unloading of the ship.

Loading and unloading costs are included in the shipping rate charged per ton, so these shipping service rates will vary according to the shipment terms under which a cargo is shipped. A higher rate per ton should be charged when the shipowner is responsible for loading and unloading costs, and a lower rate should be charged when the charterer pays for all or part of these expenses.

Recent researchers,[26] upon the empirical and analytical comparison of the three terms of shipments, indicate that most shipments of agricultural product are shipped under free-in-and-out terms where the charterer assumes responsibilities for loading and unloading costs of the ship. It appears that, in general, shippers would rather take the charges of loading and unloading of a ship than incur an increase in shipping rates by letting shipowner assume all or part of this responsibility. An important reason for this may be that many shippers own facilities for loading and unloading, thus they are able to load and unload cargoes at lower cost than shipowner would incur for hiring it done.[26]

While most shipments go under f.i.o. terms today, free-discharge terms also remain common for shipments originating from certain countries such as Argentina and South Africa. Shipment under gross terms used to be the customary procedure in world agricultural trade. However, it has been increasingly replaced by free-in-and-out and free-discharge terms. Today most shipments moved under gross terms go to underdeveloped regions in Africa and Asia. This is because when cargo handling facilities are poorly developed, as they are in most developing and underdeveloped countries, shippers appear to be more likely to let shipowners assume the responsibility for loading and unloading charges. These charges are usually higher than in other cases because mostly lighters, that are expensive, have to be used to assist the inadequate port facilities in those regions to unload a ship.[26]

SUMMARY

Rice is an important crop for the U.S. in international trade. After World War II, U.S. rice industry was facing a problem of maintaining market outlets for a production almost double the level of the prewar period. In 1973, rice acreage control programs were lifted and, as a result, the U.S. rice production has increased substantially. With this expansion of rice production in both the U.S. and foreign countries, there emerged more serious competition in the international rice market; U.S. rice exports decreased accordingly. The southern region is the most important area for both U.S. rice production and rice exports. Transportation cost for carrying rice from the U.S. southern region to world markets has been one of the major factors affecting the competitive position of U.S. rice in world markets. U.S. fleet vessels can be viewed as two types: U.S.-flag vessels and foreign-flag vessels. There has been a large difference in transportation rates between the two, with the U.S.-flag vessels operating at much higher costs. The most important way for U.S.-flag ships to continue operating and competing in the world market has been through government subsidization.

There are three major types of vessels for U.S. rice exports: liner, tanker and tramp. Tramp vessels are the most significant transportation means for carrying U.S. rice to world markets. They accounted for the largest part of U.S. rice shipments. Tankers are the least important vessels for U.S. rice exports.

There are four types of charters for cargo vessels: voyage charters, contracts of affreightment, time charters and bareboat charters. They reflect differences in terms of shipowner's involvement in the shipping operation and the division of shipping cost, etc. Voyage charter is the most frequently used arrangement.

There are two categories of terms of shipments in the world market. The first category is associated with the contractual arrangement between exporters and importers as to who is responsible for bearing the risks and costs of the trade. The second category refers to the contractual arrangement specified in a ship charter as to who, shipowner or charterers, is responsible for the loading and unloading costs of a ship, with f.i.o. (free-in-and-out) being the most preferred term in trade shipment where the charterer assumes responsibility for both loading and discharge expenses.

NOTES

1. Makasiri Sangsiri. *An Economic Analysis of Factors Affecting United States Rice Exports.* Mississippi State University, August 1986.
2. U.S.D.A. *Agricultural Statistics.* Various selected issues. 1970-1994.
3. U.S.D.A. *Rice,Outlook and Situation.* Various selected Issues.
4. Shelby Holder and Warren R. Grant. *U.S. Rice Industry.* Economics Statistics, and Cooperatives Service, USDA, 1979.
5. J.Norman Efferson. *The Production and Marketing of Rice,* Simmons Press, New Orleans.
6. Shelby H.Holder, Jr., Dale L. Shaw, and James C. Synder. *A System Model of the U.S. Rice Industry.* Economic Research Service, Technical Bulletin No. 1453, USDA. Nov. 1971.
7. U.S.D.A. *Analysis of the effects of Cost-of-Service Transportation Rates on the U.S. Grain Marketing System.* Technical Bulletin, No. 1484.
8. Earl A. Stennis, and Musa Pinar. *Analysis of Ocean Transportation Cost in International Cotton Marketing.* Staff Pater Series, #57. Department of Agricultural Economics, Mississippi State University.
9. Alberta H. Charney, Nancy D. Sidhu and John F. Due. *Short Run Cost Functions for Class II Railroads.* College of Commerce and Business Administration, University of Illinois at Urbana-Champaign.
10. David E. Moser and Michael W. Woolverton. *Estimating Barge Transport Costs for Grain and Fertilizer.* Research Bulletin 1028. University of Missouri-Columbia College of Agriculture.
11. John w. Sharp and Hugh J. McDonald. *The Impact of Vessel Size on and Optimal System of U.S. Grain Export Facilities.* Ohio Agricultural Research and Development Center. Research Bulletin, No. 1048. Oct. 1971.
12. U.S.D.A. Agricultural Marketing Service. *Cost of Watermelon Handling From Grower to Retailer.* Marketing Research Report, No.1141.
13. LeRoy Davis. *Cost of Shipping United States Grain Exports to Principal World Markets.* Department of Agricultural Economics, Iowa State University, 1968.

14. Musa Pinar. *Analysis of Ocean Transportation Costs and Tariff Barriers in International Cotton Trade.* Unpublished Ph.D. dissertation. Mississippi State University, December 1983.

15. Andrzej Olechowski and A.J.Yeats. Hidden Preference for Developing Countries: *A Note on the U.S. Import Valuation Procedure.* Quarterly Review of Economics and Business. Volume 19:3, Autumn 1979.

16. James K. Binkley and Bruce Harrer. *Major Determinants of Ocean Freight Rates for Grains: An Econometrics Analysis.* American Journal of Agricultural Economics, 1981.

17. Congress of the United States, Office of Technology Assessment. *An Assessment of Maritime Trade and Technology.* Oct. 1983.

18. Bureau of Accounts, Interstate Commerce Commission. *Rail Carload Cost Scales.* 1974.

19. John B. Harrer. *Ocean Freight Rates and Agricultural Trade.* Unpublished thesis, Purdue University, June 1979.

20. Martin Stapford. *Maritime Economics.* 1988

21. L.K. Glaser. Provision of the Food Security Act of 1985. Agriculture Information Bulletin No. 498, ERS, USDA. April 1986.

22. United Nations. *Freight Markets and Level and Structure of Freight Rates.* Report by the Secretariat of UNCTAD, New York. 1969.

23. U.S. Department of Commerce, Bureau of the Census. *Unpublished Foreign Trade Statistics,* SA705/705IT, 1986. Bureau of the Census, U.S. Department of Commerce, Washington, D.C., 1986.

24. Alabama State Docks Department at Mobile, Alabama. *Rates, Charges and Regulations Applicable at the General Cargo Piers and Warehouse.* October 29, 1976.

25. Glen Dale Norbert Cayemberg. *An Analysis of Freight Rates and Ocean Shipping of United States Grain Exports.* Department of Agricultural Economics, Iowa State University, 1969.

26. Mary Elizabeth Revelt. *Oceanborne Grain Trade: A Descriptive Analysis of Various Components of the International Shipping Industry and Their Effects on World Grain Trade.* Purdue University. December 1980.

27. Information is based on data compiled by A & P Appledore, Inc. for the United Nations Conference on Trade and Development (UNCTAD) Secretariat in 1981.

28. "Fixture" is a standard term used to describe a chartering contract in the tramp freight market.

II

THE THEORETICAL FRAMEWORK

This chapter provides an overview of international trade and information about transportation costs. International trade is important to the U.S. economy because there are economic gains obtainable for U.S. producers and merchandisers through specialization in production and trade. The value of these economic gains is sensitive to production costs and transportation costs relative to the costs experienced by producers and merchandisers in other major supply regions.

This chapter also provides information about terms of shipment that are important in understanding and analyzing transportation costs in international trade. Finally, a transportation model is introduced in order to better analyze transportation service rates in shipping rice from the U.S. to the world market and derive optimum least-cost shipping patterns.

INTERNATIONAL TRADE

Since the time of Adam Smith in the 18th century, economists have devoted substantial attention to international trade and a large part of economic theory has been developed to deal with this subject. The economic gains from national specialization and international trade are now well recognized, and numerous economic analyses associated with world trade, such as exchange rates and government policies, have been undertaken since the concepts of trade theory were developed.

In recent years, international trade has become a dominant factor in economic growth for most industrialized countries. Many newly industrialized countries have become so through major trade growth. The United States is a major trading nation, with international markets becoming increasingly important to U.S. industry.

Theoretically, nations trade with each other for fundamentally the same reasons that individuals engage in exchange of goods, to obtain the benefits of specialization. Since nations, like individuals, are not equally suited to produce all kinds of goods they need, either because they are differently endowed or for other reasons, all would benefit if each nation specialized in what it could do best and obtained its other needs through exchange. The concept of specialization constitutes the essence of international as well as interregional trade theories.

According to Adam Smith,[1] trade between two nations is based on absolute advantage and benefits to both nations. That is, when one nation is more efficient than (or has an absolute advantage over) another in the production of one commodity but is less efficient than (or has an absolute disadvantage with respect to) the other nation in producing a second commodity, then both nations can gain by each specializing in the production of the commodity of its absolute advantage and exchanging part of its output with the other nation for the commodity of its absolute disadvantage. By this process, resources are utilized in the most efficient way and the output of both commodities will increase; both nations end up consuming more of both goods. This increase in the output of both commodities measures the gains from specialization in production available to be divided between the two nations through trade.

However, absolute advantage can explain only a small portion of international trade today, such as the trade between developed and developing countries. Most of today's world trade, especially trade among developed countries, could not be explained by absolute advantage. This shortcoming was left for David Ricardo, with the law of comparative advantage, to fully explain the basis for and the gains from trade.

In 1817, David Ricardo[2] introduced the law of comparative advantage which is one of the most important and still unchallenged laws of economics with many practical applications. This law postulates that even if one nation is less efficient than (has an absolute disadvantage over) the other nation in the production of both commodities, there is still a basis for mutually beneficial trade unless the absolute disadvantage is exactly the same or in the same proportion for the two commodities. The less efficient nation should specialize in the production of and exportation of the commodity in which its absolute disadvantage is smaller (this is the commodity of its relative or comparative advantage) and import the commodity in which its absolute disadvantage is greater (this is the commodity of its relative disadvantage compared to the other nation).

Ricardo based his law of comparative advantage on a number of simplifying assumptions, the labor theory of value is one of them. This theory states that the value or price of a commodity depends exclusively on the amount of labor going into the production of the commodity. This is not true and therefore must be rejected. Labor is neither the only factor of production nor is it homogenous, and labor is not used in the same fixed proportion in the production of all commodities.

Ricardo's principle, therefore, left the actual ratios of international exchange or international prices, and the patterns of trade among nations, undetermined. How, with trade, each nation gains by specializing in the production of the commodity of its comparative advantage and exporting some of its output in exchange for the commodity of its comparative disadvantage, were determined by comparative differences in labor costs.

Heckscher and Oblin (H-O),[3,4] developed the modern theory of international trade, a more comprehensive theory known as H-O factor-proportions theory that further explained the basis for trade. The immediate basis for trade is the existence of relative price differences caused by significant national differences in any of the price-determining forces. They also state that relative differences in factor-prices are not enough to guarantee relative differences in commodity prices which are necessary if profitable trade is to develop.

Since specialization and trade rest on price differences which may be attributed to cost differences, then what is the basis for differences in cost? In addition to factor proportions and the accompanying factor price differences, there are transport costs, economies of size, external economies as well as other trade restrictions.

SIMPLE MODELS OF RICE TRADE

Figure 2 illustrates a free trade on international commodity flows in a one-commodity (rice), two-country case. For simplicity, other relevant costs such as transportation rates, are assumed to be zero and the exchange rate is assumed to be one. In a closed economy, the interaction of domestic supply and demand determines the quantity of rice produced and the price at which it is exchanged. As indicated in panel (a) in figure 2, Da represents Country A's demand curve for rice and Sa represents Country A's supply of rice. Given these two curves in country A, an equilibrium is obtained at price level Pa where quantity demanded is equal to quantity supplied.

In panel (c), Db represents Country B's demand curve for rice and Sb represents Country B's supply curve of rice. Again, given the supply and demand curves for Country B, an equilibrium is obtained at domestic price level Pb where quantity demanded is equal to quantity supplied. As indicated in Figure 2, the market equilibrium prices of rice in question in the two countries are not equal, price in Country A (Pa) is lower than price in Country B (Pb). This is due to various relative production and/or factor endowment advantages in Country A. Based on the proposition that whenever the domestic prices are different in two countries, there is room for mutually beneficial trade and the larger the difference is in the prices, the greater the gain will be from the trade. That is, if free trade is allowed, rice will be transferred from Country A into Country B in response to the price difference.

Now, assume free trade is available. In panel (b) of figure 2, international rice trade is shown. Sx curve represents an excess supply of rice by Country A which shows the quantities that Country A has available to offer at various prices. For each price this quantity equals the divergence between the domestic supply and demand curves. At price level Pa, Country A will not have any excess supply of rice, but as world price increases, quantity supplied will increase accordingly. In panel (b), Dx curve represents an excess demand for rice by Country B, showing the quantities that Country B is willing to accept (import) at various price levels. For each price, this quantity equals the horizontal distance between internal demand and supply curves. At price level Pb, Country B will not be willing to accept any quantities for rice, but as price goes down, quantity demanded will increase.

Given the excess demand and excess supply curves in the international rice market, an equilibrium is obtained at price level Pe where Country A will have a surplus of AB amount rice to export which is equal to CD amount deficit imported by Country B. Also, AB = CD = OQe which is the net rice trade, and the area PbEPa in panel (b) is the social gain from the trade for both countries.

From this illustration, we see that through free trade between two countries, competition among all consumers in both countries causes the

The Theoretical Framework

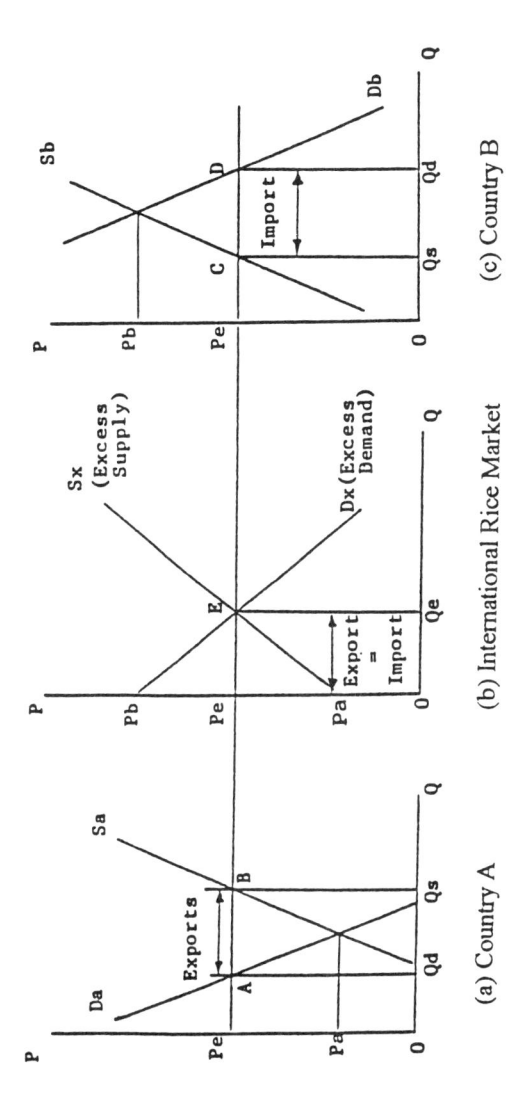

Figure 2. Free Trade Model for Rice Between Two Countries

Source: see Note 11.

domestic price of rice to increase in Country A and decrease in Country B until the combined market of the two countries is cleared. At this point, the prices of rice in both countries are equal, as is the total quantity of rice supplied for export from Country A and demanded for import in Country B. If free trade were allowed for all goods, traditional trade theory indicates that both countries would benefit through a more efficient use of their relative factor endowments and production advantages.

Now consider a more complex situation of free trade for rice in which there are two exporting countries and one importing country (figure 3). Again assume transportation costs are zero. Suppose Country A and Country C have the same relative total production and factor endowment advantage. This is graphically represented by the supply curves, Sa, Sc, with equal slopes and intercept terms in figure 3. Domestic demand for rice in both Country A and Country C need not be the same. Free trade again results in equal prices for rice across all three countries. In this case, the import demand in Country B is satisfied through exports from both Countries A and C.

TRANSPORTATION COST

Oceans and seas cover about 71 percent of planet Earth's surface. They also are the primary surfaces enabling ships to carry an estimated 70-75 percent of the volume of the world's foreign trade.[5] Ocean shipping is a growth industry due to increasing trade among countries. Transportation is requisite for trade to occur between two areas. Movement of goods generates transfer costs.

Ocean transportation service rates can be simply defined as the price (cost) paid for transferring a good from point of export to point of import and includes transfer charges and other cost items. An importing country's import prices of a certain good consist of price paid to foreign producers plus the cost of transportation. On the other hand, the local producers of the good for export receive only the domestic price of the good. The simplest way to measure this transportation cost is therefore to assume that it is met by the wastage proportion of the goods traded. This assumption implies that transportation costs are measured in terms of the exported commodity of each country.

Therefore, when Country A is exporting a certain product (say,

The Theoretical Framework

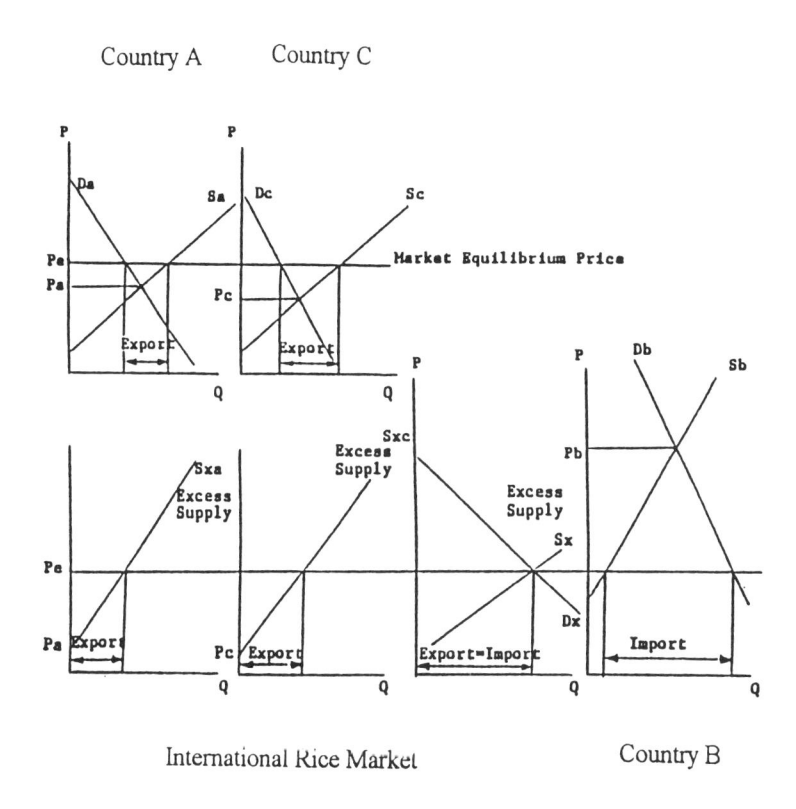

Figure 3. Free Trade Model for Rice: Three-Country Case

Source: see Note 11.

rice), the monetary value of rice reaching the importing Country B will be less than that shipped from Country A by the amount of transportation costs, the difference is "used up" as transporting costs. That is to say, the main effect of transportation costs in pure international trade is that when a certain commodity is transferred from one country to another, the monetary value received by the importing country will be less than that shipped from the exporting country by the amount of transportation costs.[6]

The implication is that the higher the transportation costs between the importing and exporting countries, the greater is the quantity of the traded commodity that will be "used up" as transport, consequently, the smaller will be the net amount of trade. In addition, transportation costs prevent the complete equalization of commodity prices that international trade would otherwise bring about.[7]

Transportation costs have been long ignored or treated in a token fashion primarily because the problem of analyzing transportation costs in international trade is a complex one. International freight rates change frequently and by large amounts relative to the commodities being shipped. For example, during the 1970's, when some production factors caused the export price of cereals to increase from $70.25 to $158.25 per metric ton (a 125 percent increase), the average rate charged for international shipments of grain went up from $8.95 to $25.72 per long ton (a 187 percent increase).[8]

Not only does the instability in the quantity of commodity demand and supply affect transportation costs, but the influence of international political policies, events, and agreements can substantially affect these costs. Cayenburg[9] examined the relationship between international events and agricultural shipping costs and found a definite correspondence between shipping rates and international events like the reopening of the Suez Canal in 1957. International agreements are institutional forces which influence shipping rates. They make the analysis of transportation costs more complex, as the rates charged under institutional forces often do not reflect the competitive situation existing in the market for transportation services. For example, the U.S. and the former U.S.S.R. made an agreement specifying that a certain amount of grain must be shipped on each other's vessels at a certain rate regardless of the rate existing in world shipping markets.[10]

Transportation cost analysis is inherently complex because many different commodities can be moved via the same transporting mode. The demand for a mode of transportation is affected by the existing demand and supply conditions for all of the commodities that can be transferred by this

The Theoretical Framework

mode at a cost less than the difference in market prices for commodities between the importing and exporting countries. This makes the pricing of transportation services a function of many variables.

To reduce the complexity of the analysis of transportation costs, this book concentrates on the international movement of rice by selected sizes of ships from port to port between the U.S. and importing countries. Although the method of determining shipping rates may vary, the rate remains a price set to cover cost of providing shipping services.

Costs encountered in owning and operating a vessel can be loosely classed under the following three categories:[9,11]

(1) Vessel ownership expenses
(2) At-sea expenses
(3) In-port expenses

Several components associated with each of these cost categories are listed below.

Vessel Ownership Expenses

A. Depreciation and Interest Depreciation is an estimate of the value of the vessels that wear out or become obsolete over the life year of the vessel. Interest is the price paid for the use of funds for ship construction, expressed as a percentage per dollar of funds borrowed.

B. Crew Wages Crew costs comprise all direct and indirect charges incurred by the crewing of vessel, including basic salaries and wages, social insurance, pensions, victuals and crew travel, and repatriation expenses. The level of manning costs is principally determined by two factors: the size of the crew and the various direct and indirect costs associated with employment.

C. Subsistence Cost of subsistence includes all the necessary living expenses which occurred during the voyage, such as supply of goods consumed by the crew.

D. Insurance Marine insurance rates depend on many factors which vary widely among ship operators, usually including ship size, trade route, company loss experience, degree of self-insurance, and numerous intangibles of management, i.e., reputation and experience of ship operator, port captain, port engineer and ship masters. A high proportion of marine insurance costs is determined by the insurance of the Hull and Machinery (H & M), which protects the owner of the vessel against physical loss or damage, and protection and indemnity (P & I) insurance, which provides coverage against third party liabilities, such as damaging a jetty or oil pollution.

E. Maintenance and repair Cost of maintenance and repair vary widely and are influenced by factors such as trade-route weather conditions, bow shape, owner's standards, and initial extra cost for reliability. This covers all outside charges associated with maintaining the vessel to the standard required by company policy on the classification society. Broadly speaking, this costs can be subdivided into three categories: routine maintenance, classification survey and breakdowns.

F. Stores and supplies Stores and supplies expenses are expenditures on consumable supplies such as spare parts, deck and engine room equipment, and lubrication oil. The most important single item is lubricating oil, since the majority of modern vessels have diesel engines and may consume several hundred liters a day while at sea. Expenditure on spare parts and replacement equipment is likely to increase with age.

G. Administration and miscellaneous expenses This category comprises ship management, communications, crew transportation, and survey fees.

H. In-port fuel A ship actively engaged in trade, whether alongside the pier, maneuvering in port, or underway, will maintain a certain minimum amount of power available. Thus there is a daily minimum for fuel required to maintain this power. This minimum power is considered as contributing to the fixed expenses of vessel ownership.

I. Return on investment Return on investment is the cost used as capital repayments for the investment in vessel purchasing or building. The rate of return on investment is predetermined, upon which the amount of the repayment is calculated.

At-sea Expenses

This category includes only one item, fuel consumed at sea, which differs from in-port fuel under the ownership costs category. In the early 1970s, when oil prices were low, little attention was paid to fuel costs in ship design, and many large vessels were fitted with turbines, since the benefits of higher power output and lower maintenance costs appeared to outweigh

their high fuel consumption.
However, when oil prices rose, the whole balance of costs changed. As a result, resources were poured into designing more fuel-efficient ships and operating practices were adjusted so that bunker consumption by the shipping industry fell sharply. In operation, the amount of fuel actually used by the ship depends on its hull condition and the speed at which it is operated.

In-port Expenses

Port charges include a wide range of fees levied against the vessel and/or cargo for the use of the facilities and services provided by the port. Charging practices vary considerably from one area to another, but broadly speaking they fall into two components, "per call" and "per day." "Per call" category includes the costs charged to the vessel each time it enters port, regardless of the actual time spent in port. These fixed costs per port call include such items as pilotage, tuggage, bunker fee, and dockage fee.

The "per day" group comprises the costs according to the number of days spent in port. The items include pier charge, fresh water, electricity, berth fee, and other service charges. While many factors can effectively influence these costs, recent research indicated that port distance, ship size and volume of trade along a particular traveling route are the most important variables in estimating the cost of shipment in rice trade.[12] The next section will therefore focus on how these factors affect shipping costs and thus freight rates charged.

EFFECT OF DISTANCE, SHIP SIZE AND TRADE VOLUME ON SHIPPING COST

Distance

Trade flow analysis has frequently based the freight rates for moving goods on the geographic distance between ports. Use of distance in this manner implies the existence of a linear relationship between distance traveled and freight rates. Most of the analysis of location is based on the premise that there exists a cost per unit of distance for moving goods. As ton-miles traveled increase, freight rates are implicitly assumed to increase proportionately at a specific cost per unit. However, Bresslen and King[13] demonstrated that most commercial transportation rates increase at a decreasing rate with distance traveled. This is particularly true in the case of ocean shipping where in-port costs represent a large percentage of the total shipping cost incurred by the carrier while at-sea costs per mile tend to decline.[14]

Distance may result in a substitution of products if the costs of overcoming the distance are greater than the value placed on obtaining the product by its purchasers. However, the transport costs over the oceans are low enough that low unit value goods like rice can be economically moved around the world by water.

The geography of land masses plays an important role in the determination of the distances for shipping goods. Because total costs increase with distance, ship operators are assumed to desire moving commodities on the shortest route possible. This may not always be true, however. While there appear more and more new technologies in the shipping industry that can reduce the effects of distance on the shipment of goods, distance is still an important factor to be considered in the analysis of transportation rates and commodity flows.[14]

Ship Size

Economies of size based on engineering principles and geometric relationships common in varying degree to many industries has long been viewed as a salient feature of modern shipping. Indeed, considerable modification of ship size through technological change has happened over the last twenty years which implies the existence of cost advantages with the use of larger ships.

Economies of size associated with vessel's ownership cost has been commonly accepted. Since shipping capacity increases faster than material requirements,[15] the capital costs and depreciation are lower per ton-mile for larger ships than for smaller ones.[16] Operating costs for ships in trade also increase at a slower rate than the proportionate rate with ship size. Manning requirements increase at a decreasing rate due to the fixed amount of tasks involved in ship operations and to increased use of automation in modern shipping.[17]

The effect of ship size on in-port cost is less clear. Port time is a major part of shipping costs, as the time spent in port represents potential opportunity costs to both shipper and ship owners. Increases in in-port costs due to the time spent in port will at some point offset decreases in ownership cost caused by larger ship size.

In other words, if either storage facilities are too small to store the amount of cargo necessary to efficiently utilize larger ships or the handling facilities are too slow to efficiently load and unload larger ships, the economic benefits of using larger ships will be reduced and may even be wiped out entirely.

The effect of economies of size on shipment costs has been supported by many researchers. Sturney[17] states that in the past 100 years, there has been a decrease in the real cost of transport and part of this is due to the use of larger ships. Increasing technology has enabled economies of size to be secured without reducing the efficiency of service.

Results of many researchers indicate that while there are economies of size in shipping, they are not without limit. Janson and Schneenson[16] point out that cargo handling operations are characterized by inherent diminishing returns attributable to ship size due to basic geometric relationships and thus place an upper limit on ship size. They found that while there are economies of ship size in at-sea operations, there are in-port diseconomies of size due to cargo handling limitations that result in larger ships spending an increasing amount of time in port than smaller ships. They conclude that overall economies of ship size are not without limit; rather, the optimal ship size is obtained by trading off at-sea economies of size with in-port diseconomies of size.

However, Robinson,[21] using empirical data from the port of Hong Kong and a regression procedure, found that larger ships spent less time in port than did smaller ships. Larger ships were also found to load and unload more quickly than smaller ships.

Heavon and Struden[19] conducted a similar study. They regressed loading time and loading rate against ship size and found that while larger vessels spent longer time in port, loading rates increased as ship size increased. They concluded that the difference between increased time spent in port and increased loading rates for larger vessels may be large enough for in-port costs to diminish as vessel size increases.

The economies of ship size were further analyzed by Kendall.[20] In his "A Theory of Optimum Ship Size," he stated that while normal trade practice was to use the largest ship available which can be accommodated at the origin and destination ports, volume of trade and length of route were also important determinants of ship size. Carriage of an increasingly large volume of product must be undertaken in order to take advantage of economies of size.

Kendall developed his theory of optimum ship size on the assumption that economies of size exist in shipping. The larger the amount of tonnage shipped on a particular route, the greater the economies of size. The optimum size of ship on a particular route was that which minimized total transport costs on that route including in-port charges. Volume of trade, value of the product, and distance travelled interacted to determine which ship size will be favored.

Bannathan and Walters[18] analyzed the relationship between at-sea and in-port expenses associated with ship size and total transportation costs. Their study indicates that increase in tonnage loaded/discharged per ship ton per day must be increasing faster than increases in daily in-port costs per ship ton if ship size is to increase. The size of vessel entering a particular port area is therefore not only directly influenced by port costs, but also by port investment decisions that affect per day load values.

It appears, however, that most grain shippers do not take much advantage of these economies of size. Perhaps restrictions on load size due to inadequate port facilities and the demand and supply conditions existing for rice are the main factors. But, from the model analyzed later in this book, even within a small range of ship sizes (25,000 to 40,000 DWT), one can see the differences in shipping costs between the larger and smaller ships with the larger ones having less cost per ton of ship.

Volume of Trade

Various researchers demonstrate that the volume of trade carried on a particular route can have a direct effect on shipping costs. Areas with large volumes of trade have relatively lower terminal costs per ton as compared to areas with smaller volumes of trade.

Lawrence[22] indicated that as volume of shipping along a particular trade route increases, it is reasonable to expect average shipping costs

The Theoretical Framework

charged to decrease. A main reason is that cargo handling facilities in ports along more active trade routes tend to be more technically advanced and more efficient. The storage availability for various types of goods, and cargo handling and ship related activities are all more likely to be better in ports along active trade routes.

Also, volumes of trade have an influence on freight rates through their impact on backhaul opportunities, where a ship hauls cargo on both ends of its voyage. In a competitive market, when a vessel carries cargo on the return end of its voyage, the rate charged for this backhaul trip is less than the rate charged on the original trip. Discharge of goods along a busy trade route increases the possibility of making a new cargo hauling contract at the unloading port on a nearby port and thereby reduces empty mileage.[11]

In general, distance of voyage, ship size and volume of trade all have certain effects on ocean freight rates. When distance of voyage increases, per unit freight rates are likely to increase but at a decreasing rate. Ship size also has significant effect on freight rate. Increases in ship size can place downward pressure on freight rates due to economies of size with respect to ownership cost and at-sea expenses. However, there are diseconomies of size relating to in-port expenses which may to a certain extent reduce or offset the economies of size. Increases in trade volumes along a particular trade route with appropriate port pricing and investment policies can cause an increase in optimum ship size along a given route. Increases in trade volumes on a specific trade route itself are also likely to put downward pressure on freight rates.

ROLE OF TRANSPORTATION COST IN INTERNATIONAL TRADE

Transportation costs of moving goods are a fundamental element of international trade. Even though transportation costs do not by themselves provide distinct bases for trade, they certainly influence the magnitude of trade flows, types of goods exchanged internationally, and the net gains and benefits from international trade. The gains from trade and specialization can accrue only if there is some means to transport goods from producing areas to consuming areas at a cost that is less than the difference in market prices between the two areas.

Therefore, transportation costs are likely to affect the degree of specialization in various countries or regions. Consequently, it may destroy the possibility of trade between the countries, or, at the least, reduce the volume of the trade. Also, transportation costs, in general, affect the prices of internationally traded goods merely because of their existence. Therefore, it is obvious that transportation costs must be included as important components of any complete explanation of national specialization and of trade patterns in the analysis of international trade.

Referring to the second section of this chapter, pure theory of international trade was explained by using two simple models of free trade, figure 2 and 3. For simplicity, transportation costs were assumed to be zero in both models. Since transportation costs do exist and are necessary for international trade, the transportation considerations should be added to the model. Figure 4 is a modified diagram of figure 2 to induce a transportation cost. In order to move rice from Country A to Country B, additional costs are incurred. Inclusion of transportation costs will cause an increase in the per unit price of rice in Country A by the amount of the transportation charges to be perceived in the international marketplace.

Since the pretrade price is lower in Country A than in Country B, if the initial differential in prices exceeds the transportation costs, there will still be a flow of exports from Country A to Country B. Figure 4 shows how the transportation costs affect international rice trade. In response to this perceived higher price, import demand for rice in Country B will decline accordingly. This is indicated by a parallel downward shift of Country B's excess demand curve from Dx to Dx' in the center panel of figure 4. The intersection of Country B's new excess demand curve Dx and Country A's excess supply curve Sx determines the amount of rice that is to be traded in the world market. Now, however, the price paid for rice is divided into two parts, the per unit F.O.B. export price and the per unit transport charge. Summation of these two prices results in the C.I.F. price in Country B. Addition of transportation costs thus drives a wedge between the price of rice in the two markets. It also results in smaller volumes of rice traded (Q'd Q's) than in the previous one (Qd Qs).

In a more complex situation in which there are two exporting countries and one importing country, as shown in figure 3, again transportation costs are added. Figure 5 illustrates the analysis of this

The Theoretical Framework 47

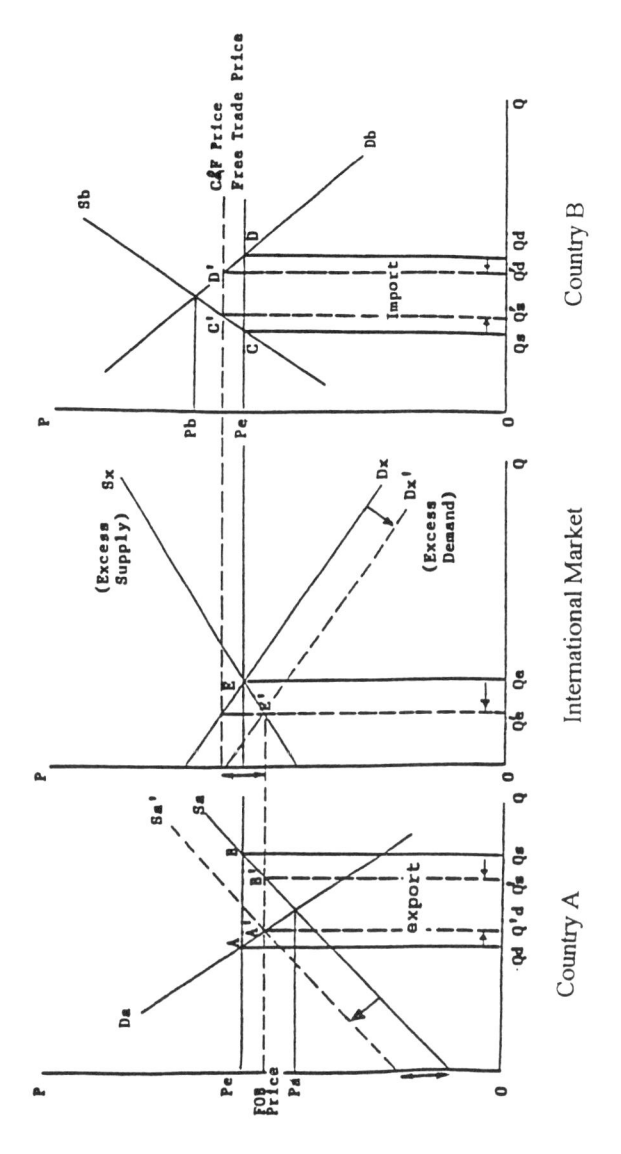

Figure 4. Effect of Transportation Cost on International Rice Market: Two-Country Case

Source: see Note 11.

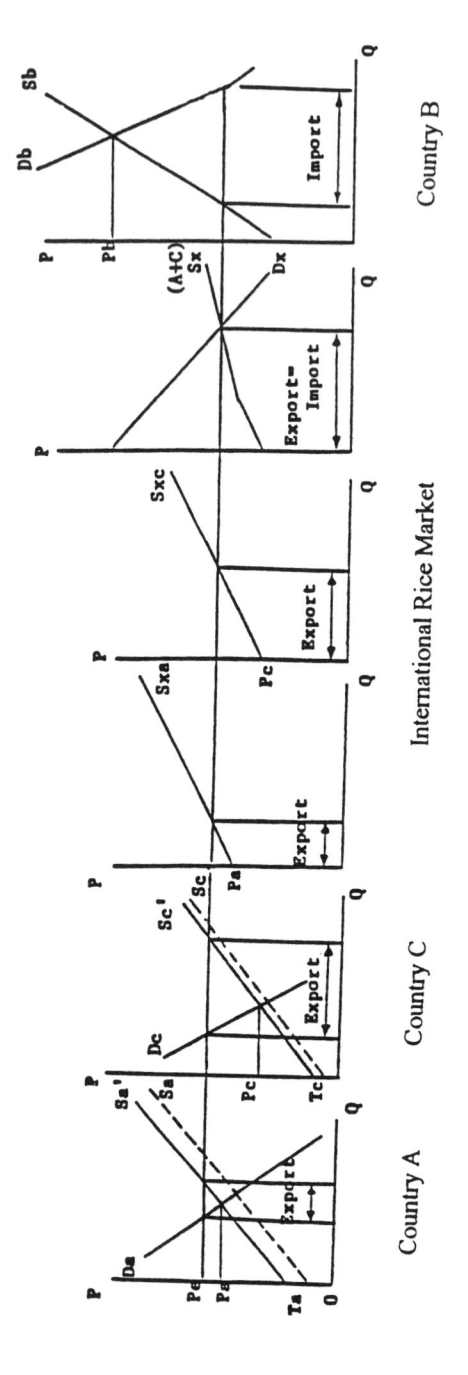

Figure 5. The Effect of Transportation Cost on International Rice Market: Three-Country Case.

Source: see Note 11.

The Theoretical Framework 49

situation. Assume Country A's per unit transport charge is higher than Country C's which might be due to a different distance of moving rice from Countries A and C to Country B. This transportation cost causes a paralled upward shift of Countries A and C's supply curves from Sa and Sc to Sa' and S'c.

The vertical distance between the new and old price intercepts is a measure of the per unit transport charge. Country A's unit transport charge (Ta) is larger than Country C's (Tc). Summation of Countries A and C's excess supply curves yields a kinked excess supply curve in the world market. The difference between the two exporting countries unit transportation costs changes the relative market share of A and C in B's importing market.

Therefore, transportation costs play a crucial role in international trade. A reduction in transport costs decreases the effects of distance on the movement of goods traded. Numerous researchers, including Beaver,[23] Sturney[17] and others, state that in the last 100 years or more, the real costs of moving commodities between countries have decreased, and it may be inferred that transport costs decreases have contributed to the increase in the volume of international trade that has occurred in recent times.

Since transportation costs cannot be eliminated, the solutions for reducing transportation costs induced trade aberrations may be stated as follows: (1) to reduce specific transportation costs as much as possible and (2) to minimize total transportation costs by obtaining an optimal flow of goods in international trade. Generally, transportation costs are not controllable by policy makers, and are essentially administered prices. However, by obtaining an optimal flow pattern of goods among exporting and importing countries, transportation costs can be reduced because buyers and sellers are free to choose markets based on free trade and its underlying assumptions.

SUMMARY

Imports and exports have become more and more important in most nation's economic activities. Nations can maximize their income by engaging in specialized production and exchange their specialized products with one another. Exchange and trade are integral to a society and are the foundation upon which economic progress is based. Today an extensive body of literature exists on the subject of international trade and its effect on the welfare of an individual country and the world as a whole. Transportation cost can significantly influence the price, volume and flow of goods traded in an international market. Conceivably, if they were high enough, transport costs could preclude trade entirely. Almost all international trade in goods is transported by sea. Thus, ocean shipping plays a central and essential role in the world economy and in world trade.

International shipping rates are therefore an important factor in world trade and improvement in the level of knowledge of what these rates are and how they are determined will facilitate more refined trade analysis. An ocean freight rate can be easily understood as the price paid for transporting a good from export point to import point. Costs encountered by shipowners can loosely be divided into three categories: ownership costs, in-port costs and at-sea costs. While many factors are known to affect transportation cost, recent research has demonstrated that transporting distance, ship size and volume of trade are the most important variables in analyzing the cost of shipping commodities throughout the world. In general, per unit freight rates are expected to increase but at a decreasing rate with increases in distance. The effect of ship size on freight rates is less clearcut. While economies of size with respect to ownership and at-sea costs are acknowledged, there are also diseconomies of size with respect to in-port costs due to diminishing returns of size associated with cargo handling operations. Increases in trade volume on a specific route are expected to put downward pressure on freight rates, both directly by decreasing ballast mileage and indirectly encouraging port efficiencies and also use of larger ships.

NOTES

1. Adam Smith. *The Wealth of Nations*. Modern Library Edition, pp. 424-426.
2. David Ricardo. *The Principles of Political Economy and Taxation*. J.M. Dent & Sons, Ltd., London, 1948
3. E. F. Heckscher. *The Effect of Foreign Trade on the Distribution of Income*. Ekonomisk Tidskrift. 1919. Reprinted in H.S. Ellis

and L.M. Metzler, Readings in the Theory of International Trade. Homewood Ill: Irwin,1950.

4. B. Ohlin. *Interregional and International Trade*. Cambridge, Mass.: Harvard University Press. 1933.

5. Congress of the United States, Office of Technology Assessment. *An Assessment of Maritime Trade and Technology*. 1983.

6. P.T. Ellsworth and J.C. Laith. *The International Economy*. MacMillan Publishing Co., Inc., New York. 1975.

7. Miltiades Chacholiodes. *Principles of International Economics*. New York: McGraw-Hill Book Company. 1981.

8. Maritime Research Incorporated. Weekly published data on ship charters, New York. 1972-1986.

9. Glen Dale Norbert Cayemberh. *An Analysis of Freight Rates and Ocean Shipping of United States Grain Exports*. Unpublished thesis, Department of Agricultural Economics, Iowa State University. 1969.

10. B.John Harrer. *Ocean Freight Rates and Agricultural Trade*. Unpublished thesis, Purdue University. June 1979.

11. Mary Elizabeth Revelt. Oceanborne Grain Trade: Adescriptive Analysis of Various Components of the International Shipping Industry and Their Effects on World Grain Trade. Unpublished thesis. Purdue University. December 1980.

12. James Binkley and Bruce Harrer. *Major Determinants of Ocean Freight Rates for Grains: An Econometrics Analysis*. American Journal of Agricultural Economics. 1981.

13. Raymend G. Bressler Jr. and Richard A. King. *Markets, Prices and International Trade*. New York: John Wiley & Sons, Inc., 1970.

14. Bulk Systems International. *Bulk Trade, Transportation and Handling*. Survey. England: McMillan House. 1980. pp.89-116.

15. The amount of material required to build a ship depends on the surface area, whereas shipping capacity depends on the volume enclosed.

16. Jan Owen Jansson and Dan Schneerson. *Economies of Scale of General Cargo Ships*. Review of Economics and Statistics, 60 (1978) pp. 287-293.

17. S. G. Sturney. *Shipping Economics*. London: MacMillan Press, Ltd., 1975.

18. Esra Bennathan and A.A. Walters. *Port Pricing and Investment Policy for Developing Countries*. New York: Oxford University Press. 1979.

19. Treavor D. Heavor and K.R. Studer. *Ship Size and Turn-Around Time: Some Empirical Evidence.* Journal of Transport Economics and Policy, 6 (1972).

20. P.M.K. Kendall. *A Theory of Optimum Ship Size.* Journal of Transport Economics and Policy, 6 (1972).

21. Roy L Nersesian. *Ships and Shipping.* Tulsa, Oklahoma. 1981.

22. S.A. Lawrence. *International Sea Transport, The Years Ahead.* Lexington, Mass.: D.C. Heath and Company. 1972.

23. S.H Beaver. *Ships and Shipping: The Geographical Consequences of Technological Progress.* Transportation Geography. McGraw-Hill Book Company. 1974.

24. UNCTAD. *Relationship Between Charges in Freight Rates and Charges in Costs of Maritime Transport and the Effect on the Export Trade of Developing Countries.* ID/B/C., 4/112, United Nations, Geneva. 1973.

III

Estimation of Shipping Cost

Ocean transportation costs are an integral part of the dada required to establish transportation models in order to find optimal shipping routes for U.S. southern region rice exports with least costs. Therefore, before the transportation models are established, it is necessary to determine the freight rates over each possible export route. The purpose of this chapter is to determine the cost per ton of shipping rice on four alternative sizes of bulk vessels from selected United States southern ports of origin to specific ports of destinations in the would under three types of flag: U.S., Japan and Liberia.

SHIPPING COST AND MEASUREMENT

The cost of running a ship depends on a combination of three factors. First, the ship itself sets the broad framework of cost through its fuel consumption, the size of crew required to operate it, and its physical condition, which dictates the requirement for repairs and maintenance. Second, running costs depend on the cost of bought-in items, such as interest rates, depreciation rates, wages paid to the crew and the level of repair costs, all of which are subject to general trends in world prices. Third, the level of costs is influenced by the efficiency with which the owner manages the operation of the ship.[1]

      Unfortunately, in shipping there is no internationally accepted standard costs classification, and this can easily lead to confusion over terminology. The approach used in this book, as described in Chapter 2, is to classify costs into three categories: ownership expenses, at-sea expenses and in-port expenses.

To determine the actual cost of shipments, it is necessary to consider all items of expenses incurred by a ship that are required for its efficient operation in accordance with existing relevant regulations and agreements. Some of the items of expense are basic to the operation of all ships, regardless of the nationality of their flag and crew, and are determining factors in whether a ship can operate at a profit. In determining cost of operation, consideration must also be given to ship design and speed. This is an implication of ship conditions as to the carrying capacity, speed, economy of operation, case of loading and unloading and crew requirement. In addition, cost per ton of shipping rice is affected by whether a vessel carries a full or partial load of rice and the ultimate destination of shipment, that is, the voyage distance and volume of trade.[2,3]

Expenses incurred at port are also important. Port location, facilities and regulations also affect shipping costs. For example, sometimes access to a port is difficult because it is located up river or because the ship must traverse dangerous channels during restricted time periods. The availability of docks, equipment, berth, labor and other facilities involved in port operations influences the costs incurred in terms of loading and discharging of cargo.[2,3] Finally, there are certain basic and uniform charges, applicable in most ports that must be considered in determining shipping cost, such as pilotage, harbor fee, towage, port fees, watchman, fresh water, electricity, warfage, handling, storage and dockage—the charge assessed on vessels occupying berths at public wharves or vessels trying to, or lying alongside any property of the port.[2,3]

There are two approaches in measuring the relationship between volume and cost: statistical approach and engineering approach. When using a statistical method to analyze a problem, a relatively large number of observations will be necessary because the larger the number of degrees of freedom, the more reliable are the results. In a statistical analysis, the input data are taken as observed, a production function is computed through use of regression techniques and the coefficients subjected to a test of significance at established confidence limits.[2]

When an engineering approach is used in analysis of a problem, the input values are determined by use of engineering data in constructing hypothetical plots, but not necessarily representative or average plots. These hypothetical plots (an ocean vessel in this case) are established for various volume ranges, and at each volume the hypothetical plot is assumed to have an optimum combination of inputs.[2]

In this book, an economic-engineering approach is used to analyze cost relationships. The vessels simulated in the analysis are selected to be as similar to rice transportation vessels in use, for each ship size and each flag, as is possible from available information.

PORTS OF ORIGIN AND PORTS OF DESTINATION

A transportation rate is to be estimated between each possible pair of locations to be considered. However, there is no set method determining regional boundaries or base points. The basic criterion is that the regions selected provide a meaningful basis for analysis of the specific problem under study. A base point is a representative point of a region or port through which all shipments would be made. Choosing a base point requires consideration of the following factors:

1. Location relative to production or consumption concentrations within the region.
2. Ocean vessel transportation facilities and the quantities usually shipped compared with other locations in the region.
3. A point through which shipments occur or might occur without overestimating or underestimating the total shipment costs to the many actual shipping points within the region.[2,4]

The original data of U.S. rice exports is obtained from the unpublished Foreign Trade Statistics Reports, SA705, SA705IT, and SA705 Supplement for the years 1981 through 1986.[5] This data set reports shipments involving more than 20 U.S. southern region exporting ports and 120 different importing countries. In order to reduce the complexity of the shipping rate information, the data is aggregated into 20 regions. This aggregation is done primarily on the basis of geographical contiguity of ports, with some attention given to the number of shipments. Number of shipments is considered in the aggregation process because of the importance of having a significant number of observations in each aggregated region.

A base point within each region is then chosen to determine the distance between each pair of origin and destination ports. The base points are selected primarily on the basis of the factors mentioned earlier in this section, the centrality of their locations with emphasis on the number of shipments.

Table 11 U.S. Southern Rice Export Regions and Base Points

Number	Base Point	Representative Ports
1	Mobile, AL	Mobile, AL Gulfport, MS Pascagoula, MS
2	Baton Rouge, LA	Baton Rouge, LA Gramercy, LA
3	New Orleans, LA	New Orleans, LA Morgan City, LA Destrehan, LA St. Rose, LA
4	Beaumont, TX	Beaumont, TX
5	Galveston, TX	Corpus Christi, TX Freeport, TX Galveston, TX Texas City, TX
6	Houston, TX	Houston, TX
7	Port Arthur, TX	Lake Charles, LA Orange, TX Port Arthur, TX Sabine, TX

Estimation of Shipping Cost

The United States southern region rice exporting ports are then grouped into seven surplus regions with each having a base point. They are: Mobile, AL, Baton Rouge, LA, New Orleans, LA, Beaumont, TX, Galveston, TX, Houston, TX, and Port Arthur, TX. These ports are shown in table 11. Countries importing rice from the United States are grouped into 13 deficit regions and a base point is chosen within each region through which all shipments would occur. These regions and base points are specified as follows and shown in table 12:

1. *KINGSTON, JAMAICA*—Central America and the Caribbean area (CAC)
2. *CALLAO, PERU*—Western South America (WSA)
3. *RIO DE JANEIRO, BRAZIL*—Eastern South America (ESA)
4. *ROTTERDAM, NETHERLANDS*—Northwestern and Central Europe (NCE)
5. *VALENCIA, SPAIN*—Northeastern and Southern Europe (NSE)
6. *HONG KONG*—Eastern Asia (EA)
7. *AQABA, JORDAN*—Western Asia (WA)
8. *SINGAPORE*—Southern and Southeastern Asia (SSA)
9. *WELLINGTON, NEW ZEALAND*—Australia and Oceania (AO)
10. *MONROVIA, LIBERIA*—Western Africa (WAF)
11. *ALGIERS, ALGERIA*—Northern Africa (NAF)
12. *MAPUTO, MOZAMBIQUE*—Eastern Africa (EAF)
13. *CAPE TOWN, SOUTH AFRICA*—Southern Africa (SAF)

LAY DAYS AND DAYS AT SEA

Lay days are the number of days a ship spends in port, as agreed to by a ship owner and a ship charterer in a ship charter. Port time has a significant influence on shipping cost. Thus, the number of days a ship spends in port is expected to have a large effect on the total costs of the voyage. There is no standard number of days that a particular type of vessel spends in a particular port. The variables involved in each individual loading and unloading are numerous and different. The provisions for port activities included under lay days vary from charter to charter.

Table 12 Countries and Regions Importing Rice from the United States

Number	Region	Base Point	Countries Represented
(1)	Central America and Caribbean (CAC)	Kingston (Jamaica)	Antilles Is. Barbados Bahamas Belize Cayman Costa Rica Dominican Republic El Salvador Guatemala Haiti Honduras Jamaica Leeward Is. Martinique Nicaragua Panama Trinidad &Tobago Windward Is.
(2)	Western & South America (WSA)	Callao (Peru)	Chile Ecuador, Peru
(3)	Eastern & South America (ESA)	Rio De Janeiro (Brazil)	Brazil Guiana
(4)	Northwestern & Central Europe (NCE)	Rotterdam (Netherland)	Belgium Denmark Finland France Netherland Norway Sweden Switzerland United Kingdom West Germany

(continued on next page)

Estimation of Shipping Cost

(5)	Northeastern & Southern Europe (NSE)	Valencia (Spain)	Gibraltar Greece Italy Malta Poland Portugal Spain Yugoslavia
(6)	Eastern Asia (EA)	Hong Kong	China(P.R.C) Hong Kong Japan South Korea Taiwan
(7)	Western Asia (WA)	Aqaba (Jordan)	Bahrain Cyprus Jordan Iran Israel Kuwait Lebanon Oman Qatar Saudi Arabia Turkey Yemen
(8)	Southern & Southeastern Asia (SSA)	Singapore (Singapore)	Bangladesh India Indonesia Philippines Singapore
(9)	Australia & Oceania (AO)	Wellington (New Zealand)	Australia New Zealand

(continued on next page)

(10)	Western Africa (WAF)	Monrovia (Liberia)	Angola Benin Cameroon Canary Congo Gabon Gambia Ghana Guinea Guinea-Bissau Ivory Coast liberia Mouritius Nigeria Senegal Sierra Leone Togo Zaire
(11)	Northern Africa (NAF)	Algiers (Algeria)	Algeria Egypt Libya Morocco Sudan Tunisia
(12)	Eastern Africa (EAF)	Maputo (Mozambique)	Afars-Issas Ethiopia Kenya Malagasy Mozambique Seychelles Somalia Tanzania
(13)	Southern Africa (SAF)	Cape Town (Rep. of S. Africa)	Republic of South Africa

For, example, in some charters, lay days include time for loading and unloading of the cargo, while in other ship charters lay days may apply only to the time needed for mechanical repairs.[6] Also, a 30,000 DWT vessel might load in four days; another identical ship could conceivably require twice that amount of time. This is understandable because, first, a vessel's arrival in port does not necessarily mean that the ship is ready to receive the cargo or that the elevator is ready to deliver rice to the vessel. By the same token, a vessel that has finished loading can conceivably remain in port for several days awaiting final sailing orders. Secondly, availability of a berth at the elevator and prevailing weather conditions during loading period vary from time to time. Thirdly, seasonality of rice harvest also affects time in port. Finally, different ports have different facilities and different loading and unloading capacities.

The time spent in port is generally used for passing customs and quarantine, inspection of rice holds for suitability of loading and moving to and from berth. In this book, information on average days spent in port in major U.S. southern ports was obtained by contacting port authorities and the Maritime Administration, U.S. Department of Transportation. Average days spent in major U.S. southern ports by different size ships are listed in tables 13 through 15.

Days spent at sea by a ship are important in calculating at-sea expenses, and thus can affect shipping costs substantially. Days at sea can be calculated for certain routes by dividing the distance between ports by the number of miles each ship travels during one day. Therefore, distance between each pair of origin and destination ports needs to be determined first. There are generally three major approaches for measuring the distance between two points. The first one is to use actual distance, which is the most accurate approach. The second method is to use air miles in lieu of actual miles in the development of regression estimates of transportation costs. Estimates of air miles can be easily obtained by use of spherical geometry and a computer program. The third approach is to estimate distance between two points by measuring the map distance between the two points.[7]

In this book, the first approach is used to measure the actual distance between each pair of exporting and importing ports. Those distances are taken from Reed's Marine Distance Tables.[8]

After the distances between each pair of ports are determined, days a ship spends at sea are calculated by dividing the distance between ports by the number of miles the ship travels during one day. In this book, information shows that the ships used for analysis travel at 16 knots.[9] This means that they can sail 384 nautical miles[10] a day. Distances between each pair of points or relevant ports, and the calculated days spent at sea by different size ships are shown in tables 13 through 15.

Table 13 Days in Port, Days at Sea, and Distance Between Export Points
(for 25,000 and 30,000 DWT ships)

Origin Destination	(1)	(2)	(3)	(4)	(5)	(6)	(7)
		(Distance is in 000 nautical miles)					
1. (CAC) Kingston (Jamaica)							
Distance	1.12	1.29	1.16	1.27	1.26	1.30	1.30
Days at sea	3	3.4	3	3.4	3.3	3.4	3.4
Days in port	4	4	4	7	3	5	4
2. (WSA) Callao (Peru)							
Distance	2.76	2.90	2.77	3.06	2.87	2.92	2.94
Days at sea	8.2	8.6	8.2	9.0	8.5	8.6	8.7
Days in port	4	4	4	7	3	5	4
3. (ESA) Rio de Janeiro (Brazil)							
Distance	5.08	5.27	5.14	5.51	5.32	5.37	5.40
Days at sea	13.3	13.8	13.4	14.4	13.9	14.0	14.1
Days in port	4	4	4	7	3	5	4

(continued on next page)

4. (NCE) Rotterdam (Netherlands)							
Distance	4.83	5.01	4.88	5.26	5.07	5.12	5.15
Days at sea	12.6	13.1	12.7	13.7	13.2	13.4	13.4
Days in port	4	4	4	7	3	5	4

5. (NSE) Valencia (Spain)							
Distance	4.95	5.13	5.00	5.38	5.19	5.24	5.27
Days at sea	12.9	13.4	13.1	14.0	13.6	13.7	13.8
Days in port	4	4	4	7	3	5	4

6. (WA) Aqaba (Jordan)							
Distance	6.80	7.01	6.87	7.22	7.03	7.07	7.10
Days at sea	18.7	19.3	19.0	19.8	19.3	19.0	19.5
Days in port	4	4	4	7	3	5	4

7. (SSA) Singapore (Singapore)							
Distance	11.5	11.7	11.5	11.9	11.7	11.8	11.8
Days at sea	31.0	31.4	31.0	32.0	31.5	31.6	31.7
Days in port	4	4	4	7	3	5	4

8. (EA) Hong Kong							
Distance	10.6	10.8	10.6	10.9	10.7	10.8	10.8
Days at sea	28.6	29.1	28.7	29.5	29.0	29.1	29.2
Days in port	4	4	4	7	3	5	4

(continued on next page)

9. (AO) Wellington (New Zealand)							
Distance	7.92	8.07	7.94	8.23	8.04	8.09	8.12
Days at se	21.7	22.2	21.7	22.5	22.0	22.1	22.2
Days in port	4	4	4	7	3	5	4

10. (WAF) Monrovia (Liberia)							
Distance	4.83	5.01	4.88	5.25	5.06	5.11	5.14
Days at sea	12.6	13.1	12.7	13.7	13.2	13.3	13.4
Days in port	4	4	4	7	3	5	4

11. (NAF) Algiers (Algeria)							
Distance	4.97	5.15	5.02	5.40	5.21	5.26	5.29
Days at sea	13.0	13.0	13.1	14.1	13.6	13.7	13.8
Days in port	4	4	4	7	3	5	4

12. (EAF) Maputo (Mozambique)							
Distance	9.11	9.26	9.13	9.44	9.25	9.30	0.33
Days at sea	24.8	25.2	24.8	25.6	25.1	25.3	25.3
Days in port	4	4	4	7	3	5	4

13. (SAF) Cape Town (S. Africa)							
Distance	7.24	7.42	7.29	7.66	7.47	7.52	7.55
Days at sea	18.9	19.4	19.0	20.0	19.5	19.6	19.7
Days in port	4	4	4	7	3	5	4

Note: Column headings are as follows: (1) Mobile; (2) Baton Rouge; (3) New Orleans; (4) Beaumont; (5) Galveston; (6) Houston; and (7) Port Arthur.

Table 14 Days in Port, Days at Sea, and Distance Between Export Points (for 35,000 DWT ships)

Origin Destination	(1)	(2)	(3)	(4)	(5)	(6)	(7)
		(Distance is in 000 nautical miles)					
1. (CAC) Kingston (Jamaica)							
Distance	1.12	1.29	1.16	1.27	1.26	1.30	1.30
Days at sea	3.0	3.4	3.0	3.4	3.3	3.4	3.4
Days in port	4.5	4.5	4.5	7.5	3.5	5.5	4.5
2. (WSA) Callao (Peru)							
Distance	2.76	2.90	2.77	3.06	2.87	2.92	2.94
Days at sea	8.2	8.6	8.2	9.0	8.5	8.6	8.7
Days in port	4.5	4.5	4.5	7.5	3.0	5.5	4.5
3. (ESA) Rio de Janeiro (Brazil)							
Distance	5.08	5.27	5.14	5.51	5.32	5.37	5.40
Days at sea	13.3	13.8	13.4	14.4	13.9	14.0	14.1
Days in port	4.5	4.5	4.5	7.5	3.5	5.5	4.5
4. (NCE) Rotterdam (Netherlands)							
Distance	4.83	5.01	4.88	5.26	5.07	5.12	5.15
Days at sea	12.6	13.1	12.7	13.7	13.2	13.4	13.4
Days in port	4.5	4.5	4.5	7.5	3.5	5.5	4.5

(continued on next page)

5. (NSE) Valencia (Spain)							
Distance	4.95	5.13	5.00	5.38	5.19	5.24	5.27
Days at sea	12.9	13.4	13.1	14.0	13.6	13.7	13.8
Days in port	4.4	4.5	4.5	7.5	3.5	5.5	4.5

6. (WA) Aqaba (Jordan)							
Distance	6.80	7.01	6.87	7.22	7.03	7.07	7.10
Days at sea	18.7	19.3	19.0	19.8	19.3	19.0	19.5
Days in port	4.5	4.5	4.5	7.5	3.5	5.5	4.5

7. (SSA) Singapore (Singapore)							
Distance	11.5	11.7	11.5	11.9	11.7	11.8	11.8
Days at sea	31.0	31.4	31.0	32.0	31.5	31.6	31.7
Days in port	4.5	4.5	4.5	7.5	3.5	5.5	4.5

8. (EA) Hong Kong							
Distance	10.6	10.8	10.6	10.9	10.7	10.8	10.8
Days at sea	28.6	29.1	28.7	29.5	29.0	29.1	29.2
Days in port	4.5	4.5	4.5	7.5	3.5	5.5	4.5

9. (AO) Wellington (New Zealand)							
Distance	7.92	8.07	7.94	8.23	8.04	8.09	8.12
Days at sea	21.7	22.2	21.7	22.5	22.0	22.1	22.2
Days in port	4.5	4.5	4.5	7.5	3.5	5.5	4.5

(continued on next page)

10. (WAF) Monrovia (Liberia)							
Distance	4.83	5.01	4.88	5.25	5.06	5.11	5.14
Days at sea	12.6	13.1	12.7	13.7	13.2	13.3	13.4
Days in port	4.5	4.5	4.5	7.5	3.5	5.5	4.5
11. (NAF) Algiers (Algeria)							
Distance	4.97	5.15	5.02	5.40	5.21	5.26	5.29
Days at sea	13.0	13.0	13.0	14.1	13.6	13.7	13.8
Days in port	4.5	4.5	4.5	7.5	3.5	5.5	4.5
12. (EAF) Maputo (Mozambique)							
Distance	9.11	9.26	9.13	9.44	9.25	9.30	9.33
Days at sea	24.8	25.2	24.8	25.6	25.1	25.3	25.3
Days in port	4.5	4.5	4.5	7.5	3.5	5.5	4.5
13. (SAF) Cape Town (S. Africa)							
Distance	7.24	7.42	7.29	7.66	7.47	7.52	7.55
Days at sea	18.9	19.4	19.0	20.0	19.5	19.6	19.7
Days in port	4.5	4.5	4.5	7.5	3.5	5.5	4.5

Note: Column headings are as follows: (1) Mobile; (2) Baton Rouge; (3) New Orleans; (4) Beaumont; (5) Galveston; (6) Houston; and (7) Port Arthur.

Table 15 Days in Port, Days at Sea, and Distance Between Export Points
(for 40,000 DWT ships)

Origin Destination	(1)	(2)	(3)	(4)	(5)	(6)	(7)
	\multicolumn{7}{c}{(Distance is in 000 nautical miles)}						
1. (CAC) Kingston (Jamaica)							
Distance	1.12	1.29	1.16	1.27	1.26	1.30	1.30
Days at sea	3.0	3.4	3.0	3.4	3.3	3.4	3.4
Days in port	4.5	4.5	4.5	8.0	3.5	5.5	4.5
2. (WSA) Callao (Peru)							
Distance	2.76	2.90	2.77	3.06	2.87	2.92	2.94
Days at sea	8.2	8.6	8.2	9.0	8.5	8.6	8.7
Days in port	4.5	4.5	4.5	8.0	3.5	5.5	4.5
3. (ESA) Rio de Janeiro (Brazil)							
Distance	5.08	5.27	5.14	5.51	5.32	5.37	5.40
Days at sea	13.3	13.8	13.4	14.4	13.9	14.0	14.1
Days in port	4.5	4.5	4.5	8.0	3.5	5.5	4.5
4. (NCE) Rotterdam (Netherlands)							
Distance	4.83	5.01	4.88	5.26	5.07	5.12	5.15
Days at sea	12.6	13.1	12.7	13.7	13.2	13.4	13.4
Days in port	4.5	4.5	4.5	8.0	3.5	5.5	4.5

5. (NSE) Valencia (Spain)

Distance	4.95	5.13	5.00	5.38	5.19	5.24	5.27
Days at sea	12.9	13.4	13.1	14.0	13.6	13.7	13.8
Days in port	4.4	4.5	4.5	8.0	3.5	5.5	4.5

6. (WA) Aqaba (Jordan)

Distance	6.80	7.01	6.87	7.22	7.03	7.07	7.10
Days at sea	18.7	19.3	19.0	19.8	19.3	19.0	19.5
Days in port	4.5	4.5	4.5	8.0	3.5	5.5	4.5

7. (SSA) Singapore (Singapore)

Distance	11.5	11.7	11.5	11.9	11.7	11.8	11.8
Days at sea	31.0	31.4	31.0	32.0	31.5	31.6	31.7
Days in port	4.5	4.5	4.5	8.0	3.5	5.5	4.5

8. (EA) Hong Kong

Distance	10.6	10.8	10.6	10.9	10.7	10.8	10.8
Days at sea	28.6	29.1	28.7	29.5	29.0	29.1	29.2
Days in port	4.5	4.5	4.5	8.0	3.5	5.5	4.5

9. (AO) Wellington (New Zealand)

Distance	7.92	8.07	7.94	8.23	8.04	8.09	8.12
Days at sea	21.7	22.2	21.7	22.5	22.0	22.1	22.2
Days in port	4.5	4.5	4.5	8.0	3.5	5.5	4.5

(continued on next page)

10. (WAF) Monrovia (Liberia)							
Distance	4.83	5.01	4.88	5.25	5.06	5.11	5.14
Days at sea	12.6	13.1	12.7	13.7	13.2	13.3	13.4
Days in port	4.5	4.5	4.5	8.0	3.5	5.5	4.5
11. (NAF) Algiers (Algeria)							
Distance	4.97	5.15	5.02	5.40	5.21	5.26	5.29
Days at sea	13.0	13.1	13.0	14.1	13.6	13.7	13.8
Days in port	4.5	4.5	4.5	8.0	3.5	5.5	4.5
12. (EAF) Maputo (Mozambique)							
Distance	9.11	9.26	9.13	9.44	9.25	9.30	9.33
Days at sea	24.8	25.2	24.8	25.6	25.1	25.3	25.3
Days in port	4.5	4.5	4.5	8.0	3.5	5.5	4.5
13. (SAF) Cape Town (S. Africa)							
Distance	7.24	7.42	7.29	7.66	7.47	7.52	7.55
Days at sea	18.9	19.4	19.0	20.0	19.5	19.6	19.7
Days in port	4.5	4.5	4.5	8.0	3.5	5.5	4.5

Note: Column headings are as follows: (1) Mobile; (2) Baton Rouge; (3) New Orleans; (4) Beaumont; (5) Galveston; (6) Houston; and (7) Port Arthur.

ASSUMPTIONS OF THE ESTIMATION

The costs of transporting rice (fixed and variable costs) from U.S. southern region ports to international markets are calculated on the basis of the following assumptions.[12,13,2]

1. Since there is competition between American and foreign flag vessels, the cost of transportation of rice is estimated for three different-flag ships: U.S., Japan and Liberia-flag vessels. The U.S.-flag vessels are assumed to be repaired in the U.S. and Japan and Liberia-flag ships were repaired in Asia. U.S.-flag ships are constructed in U.S., and Japanese-flag and Liberia-flag ships are constructed in Japan. The Liberia-flag vessel wage costs are based on a Taiwanese crew and the U.S.-flag and Japan-flag vessel wage costs assume crews composed of all nationals. Therefore, the shipping costs estimated are expected to have large differences.

2. Four vessel sizes are selected in this analysis for shipping rice from U.S. southern region ports to world markets. They are 25,000 DWT,[14] 30,000 DWT, 35,000 DWT and 40,000 DWT. Information from Maritime Administration, U.S. Department of Transportation indicates that these four ship sizes are "ideal" representatives of the bulk carriers that have been most commonly used for shipping rice in the world trade. The transportation costs for shipping rice from U.S. southern ports to importing countries are calculated for each of the four ship sizes so that it is easy to compare the shipping costs under different vessel sizes. As mentioned earlier, ship size has direct effect on shipping cost. In most situations, the rates per unit charged on smaller ships are higher than the rates charged on larger ships.

3. All of the four ships are assumed to utilize at least 95 percent of available cargo space outbound when shipping rice and to obtain 60 percent of a normal full load on the return trip.

4. The ships under estimation are assumed to have a 20 year useful life (new construction) and a mean age of 10 years. Scrap value at the end of the 20-year period is 2.5 percent of total construction cost. Return on investment is assumed to be 12 percent on the valuation of the ship.

5. The ships under study are constructed solely with borrowed capital and 12 percent interest is payable annually on the undepreciated investment of the vessel.

6. Ships under U.S.-flag are constructed without the aid of federal construction differential subsidy.

7. Average voyage days per year for each ship for commercial operation are assumed to be 350 days with an expected 15 days per year required for vessel lay-up and repairs.

8. Estimation of shipping costs is based on the distance between ports, the number of days at sea and number of days in port loading for each ship size.

THE ESTIMATION OF SHIPPING COST

As stated before, transportation costs of shipping rice from U.S. ports to importing countries can be classified into three categories: vessel ownership expenses, at-sea expenses and in-port expenses. The formulas and cost items used to calculate and estimate vessel ownership, at-sea and in-port expenses are based on information obtained from the Maritime Administration, U.S. Department of Transportation,[15] "Maritime Transportation of Unified Cargo: A Comparative Economic Analysis of Break-Bulk and Unified Load Systems,"[16] and the thesis entitled "Cost of Shipping United States Grain Exports to Principal World Markets" by M. LeRoy Davis.[2] The physical characteristics of ships used in this analysis are shown in table 16.

Table 16 Characteristics of Bulk Rice Vessels by Different Flag and Size

DWT Flag	25,000 DWT U.S.	Japan	Liberia	30,000 DWT U.S.	Japan	Liberia
Construction cost ($ million)	29.81	11.25	11.25	35.78	13.50	13.50
Crew size	26	21	20	26	21	20
Speed at sea (in knots)	16	16	16	16	16	16
Fuel consumption (tons/day)						
1. at sea	50	50	50	50	50	50
2. in port	5	5	5	5	5	5
Powered by	Diesel	Diesel	Diesel	Diesel	Diesel	Diesel
Cargo capacity	23,750	23,750	23,750	28,500	28,500	28,500

(continued on next page)

Estimation of Shipping Cost 73

DWT	35,000 DWT			40,000 DWT		
Flag	U.S.	Japan	Liberia	U.S.	Japan	Liberia
Construction cost						
($ million)	41.74	15.75	15.75	47.70	18.00	18.00
Crew size	26	21	20	26	21	20
Speed at sea						
(in knots)	16	16	16	16	16	16
Fuel consumption						
(tons/day)						
1. at sea	50	50	50	50	50	50
2. in port	5	5	5	5	5	5
Powered by	Diesel	Diesel	Diesel	Diesel	Diesel	Diesel
Cargo capacity	33,250	33,250	33,250	38,000	38,000	38,000

Sources: 1. James E. Caponiti, Chief, Division of Ship Operating Cost, Maritime Administration, U.S. Department of Transportation. Private communication. 1989.
  2. Thomas Pozemski, Chief, Division of Cost Estimates and Analysis, Maritime Administration, U.S. Department of Transportation. Private communication. 1989.
  3. Office of Technology Assessment, U.S. congress, An Assessment of Maritime Trade and Technology. 1983.

Note: When more than one source was available, an approximate average was used.

Vessel Ownership Expenses

Vessel ownership expenses include items of interest and depreciation, in-port fuel, return on investment, crew wages, subsistence, store and supplies expenses, maintenance and repair, and insurance.

1. Interest and Depreciation

Since a 12 percent interest rate is assumed to be payable annually on the capital investment of ship, interest payment is calculated as:

$(0.12)(T.C.C.)$[17] for the first year, and
$(0.12)(T.C.C./20)$ for the twentieth year.

Average annual interest payment

$$= \left[\frac{(0.12 + 0.12/20)}{2}\right] (T.C.C.)$$

$$= [(0.12 + 0.006)/2] (T.C.C.)$$

$$= (0.063)(T.C.C.)$$

Under the assumption of 350 voyage days per year for a ship, average annual interest expense per voyage day

$$= \frac{(0.063)(T.C.C.)}{350 \text{ voyage days/year}}$$

$$= (0.00018)(T.C.C.)$$

For depreciation calculation, "straight-line" method is applied, i.e., the vessel is depreciated in equal amounts over its assumed 20 year life. And the scrap value is assumed to be 2.5 percent of total construction cost.

Value to be depreciated = $(100\% - 2.5\%)(T.C.C.)$

Average depreciation expense per voyage day

$$= \frac{\text{Value to be depreciated}}{(20 \text{ years})(350 \text{ voyage days/year})}$$

$$= \frac{(1-0.025)(T.C.C.)}{(20)(350)}$$

$$= \frac{(0.975)(T.C.C.)}{7000 \text{ days}}$$

Estimation of Shipping Cost 75

Thus, amortization expense per voyage day

= interest expense + depreciation expense

= (0.00018)(T.C.C.) + (0.975/7000) (T.C.C.)

= (0.00018 + 0.000139)(T.C.C.)

= (0.000319)(T.C.C.)

Interest and depreciation expenses calculated from the above derived relationship are listed in table 17 for each of the four ship sizes and by different flags.

Table 17 Vessel Ownership Expenses: Interest and Depreciation

Size	25,000 DWT			30,000 DWT		
Flag	U.S.	Japan	Liberia	U.S.	Japan	liberia
Expense*	9,509	3,589	3,589	11,414	4,307	4,3071

Size	35,000 DWT			40,000 DWT		
Flag	U.S.	Japan	Liberia	U.S.	Japan	Liberia
Expense*	13,315	5,024	5,024	15,216	5,742	5,742

*Expenses are in *dollar value* per voyage day.

2. In-port Fuel

In-port fuel consumption for different size ships is listed in table 16. Information on fuel prices is obtained from The Journal of Commerce, "International Bunker Prices," and from contacting some port authorities. Fuel costs range from $155 to $164 per metric ton for marine diesel oil (MDO) which is the most popularly used fuel for maritime vessels. The average price for MDO is therefore calculated as $160 per metric ton which is equivalent to $145 per ton. In-port fuel expense is then calculated for vessels of different sizes and flags on the above basis which is $726 per voyage day for vessels of all sizes and all flags.

3. Return on Investment

On the basis of assumptions stated before, return on investment is 12 percent after taxes on the valuation of the ship. Vessels have a twenty year useful life and a mean age of ten years. Return on investment is calculated as follows:

$$2\{12\% \frac{[(T.C.C.-8/20 T.C.C.)+(T.C.C.-9/20 T.C.C.)+(T.C.C.-10/20\ T.C.C.)]}{3}\}$$

$$= 2[12\% \frac{(3\ T.C.C. - 27/20\ T.C.C.)}{3}]$$

$$= 2[12\% (T.C.C. - 9/20\ T.C.C.)]$$

$$= 2[12\% (11/20\ T.C.C.)]$$

$$= 2\ (33/500\ T.C.C.)$$

$$= 0.132\ (T.C.C.)$$

Return on investment per voyage day

$$= \frac{0.132\ (T.C.C.)}{350\ \text{voyage days/ year}}$$

Return on investment for vessels of different sizes and different flags is calculated on the above basis and listed in table 18.

Estimation of Shipping Cost

Table 18 Vessel Ownership Expenses: Return on Investment

Size	25,000 DWT			30,000 DWT		
Flag	U.S.	Japan	Liberia	U.S.	Japan	Liberia
Expense[*]	11,242	4,242	4,242	13,494	5,091	5,091

Size	35,000 DWT			40,000 DWT		
Flag	U.S.	Japan	Liberia	U.S.	Japan	Liberia
Expense[*]	15,741	5,940	5,940	17,989	6,788	6,788

[*]Expenses are in *dollar value* per voyage day.

    4. *Crew Wages and Allowance*

    5. *Subsistence*

    6. *Store and Supplies Expenses*

    7. *Maintenance and Repair*

    8. *Insurance*

Information on these five types of expenses on different size and flag vessels is obtained from Maritime Administration, U.S. Department of Transportation and is listed in table 19.

Total vessel ownership expenses include the above eight items and are listed in table 20.

Table 19 Five Other Vessel Ownership Expenses

Size - Flag	(1)	(2)	(3)	(4)	(5)
		(dollars per voyage day)			
25,000 DWT					
U.S.-flag	6,700	300	800	2,000	1,600
Japan-flag	5,100	200	300	700	800
Liberia-flag	1,400	100	300	700	700
30,000 DWT					
U.S.-flag	6,700	300	800	2,000	1,600
Japan-flag	5,100	200	300	700	800
Liberia-flag	1,400	100	300	700	700
35,000 DWT					
U.S.-flag	6,700	300	800	2,000	1,600
Japan-flag	5,100	200	300	700	800
Liberia-flag	1,400	100	300	700	700
40,000 DWT					
U.S.-flag	6,700	300	800	2,000	1,600
Japan-flag	5,100	200	300	700	800
Liberia-flag	1,400	100	300	700	700

Source: James E. Caponiti, Chief, Division of Ship Operating Cost, Maritime Administration, U.S. Department of Transportation. Private conversation, 1989.

Notes: Column headings are as follows: (1) Wages and allowance; (2) Subsistence; (3) Stores and supplies; (4) Maintenance and repair; and (5) Insurance.

Estimation of Shipping Cost

Table 20 Total Vessel Ownership Expenses

Size Flag	25,000 DWT U.S.	25,000 DWT Japan	25,000 DWT Liberia	30,000 DWT U.S.	30,000 DWT Japan	30,000 DWT Liberia
Interest and depreciation	9,505	3,589	3,589	11,414	4,307	4,307
Wages and allowance	6,700	5,100	1,400	6,700	5,100	1,400
Subsistence	300	200	100	300	200	100
Stores and supplies	800	300	300	800	300	300
Maintenance and repair	2,000	700	700	2,000	700	700
Insurance	1,600	800	700	1,600	800	700
In-port fuel	726	726	726	726	726	726
Return on investment	11,242	4,242	4,242	13,494	5,091	5,091
Total	32,877	15,657	11,757	37,034	17,224	13,324

Size Flag	35,000 DWT U.S.	35,000 DWT Japan	35,000 DWT Liberia	40,000 DWT U.S.	40,000 DWT Japan	40,000 DWT Liberia
Interest and depreciation	13,315	5,024	5,024	15,216	5,742	5,742
Wages and allowance	6,700	5.100	1,400	6,700	5,100	1,400
Subsistence	300	200	100	300	200	100
Stores and supplies	800	300	300	800	300	300
Maintenance and repair	2,000	700	700	2,000	700	700
Insurance	1,600	800	700	1,600	800	700
In-port fuel	726	726	726	726	726	726
Return on investment	15,741	5,940	5,940	17,989	6,788	6,788
Total	41,182	18,790	14,890	45,331	20,356	16,456

Note: Expenses are in dollar value per voyage day.

At-sea Expenses

This category contains only at-sea fuel expense. At-sea fuel consumption for different size ships is listed in table 16. Price of MDO (Marine Diesel Oil) is obtained from The Journal of Commerce, which is $160 per metric ton, or $145 per ton. In this analysis, the value of $145 per ton is used. The in-port fuel expense has been deducted from the total amount of fuel expense. At-sea expenses are then calculated as $6,524 for all twelve ships.

In-port Expenses

As stated in Chapter II, in-port charges are expenditures incurred during the time a ship stays in port. They are charges levied on the vessel for the general use of port facilities, including dockage and wharfage charges, and the provision of the basic port infrastructure. They also cover the various services that the vessel uses in port. These expenses are subdivided into two groups "per call" charge and "per day" charge.

Information on port expenses for the seven major ports of origin in this analysis is obtained through communication with the authorities and other personnel in those ports. The port expenses are shown in table 21.

Table 21 Port Expenses by Vessel Size and Flag

Size		25,000 DWT			30,000 DWT	
Flag	U.S.	Japan	Liberia	U.S.	Japan	Liberia
Expense per day	350	350	350	380	380	380
Expense per call	1,030	1,030	1,030	1,245	1,245	1,245

Size		35,000 DWT			40,000 DWT	
Flag	U.S.	Japan	Liberia	U.S.	Japan	Liberia
Expense per day	440	440	440	490	490	490
Expense per call	1,300	1,300	1,300	1,520	1,520	1,520

Source: Data from private communication with port personnel and authorities in ports of Mobile, Baton Rouge, New Orleans, Beaumont, Houston, Galveston and Port Arthur, 1990,

Estimation of Shipping Cost

Based on the above information on the three types of shipping costs, the relationship of total shipping cost per voyage day per ton is the summation of three types of expenses, times the number of days used, divided by cargo tonnage. This is expressed as follows:

1.4^{18} [Vessel ownership expenses (number of voyage days)
  + At-sea expenses (number of sea days)
  + In-port expenses per day (number of days in port)
  + In-port expenses per call (number of port calls)]
  ÷ cargo tonnage

The number of voyage days is at-sea days plus days spent in port, which are both listed in tables 13 through 15 for all size ships.

Based on the above relationship and substituting values into the formulas, the total shipping cost per ton for each of the three flags and four sizes of ships is calculated as follows:

1. *25,000 DWT, U.S.-flag ship*

    T.C./ton = 1.4 [$32,877 (No. of voyage days)
      + $6,524 (No. of sea days)
      + $350 (No. of port days)
      + $1,030] / 23,750 tons

2. *25,000 DWT, Japan-flag ship*

    T.C./ton = 1.4 [$15,657 (No. of voyage days)
      + $6,524 (No. of sea days)
      + $350 (No. of port days)
      + $1,030] / 23,750 tons

3. *25,000 DWT, Liberia-flag ship*

    T.C./ton = 1.4 [$11,757 (No. of voyage days)
      + $6,524 (No. of sea days)
      + $350 (No. of port days)
      + $1,030] / 23,750 tons

4. *30,000 DWT, U.S.-flag ship*

   T.C./ton = 1.4 [$37,034 (No. of voyage days)
   + $6,524 (No. of sea days)
   + $380 (No. of port days)
   + $1,245] / 28,500 tons

5. *30,000 DWT, Japan-flag ship*

   T.C./ton = 1.4 [$17,224 (No. of voyage days)
   + $6,524 (No. of sea days)
   + $380 (No. of port days)
   + $1,245] / 28,500 tons

6. *30,000 DWT, Liberia-flag ship*

   T.C./ton = 1.4 [$13,324 (No. of voyage days)
   + $6,524 (No. of sea days)
   + $380 (No. of port days)
   + $1,245] / 28,500 tons

7. *35,000 DWT, U.S.-flag ship*

   T.C./ton = 1.4 [$41,182 (No. of voyage days)
   + $6,524 (No. of sea days)
   + $440 (No. of port days)
   + $1,300] / 33,250 tons

8. *35,000 DWT, Japan-flag ship*

   T.C./ton = 1.4 [$18,790 (No. of voyage days)
   + $6,524 (No. of sea days)
   + $440 (No. of port days)
   + $1,300] / 33,250 tons

9. *35,000 DWT, Liberia-flag ship*

   T.C./ton = 1.4 [$14,890 (No. of voyage days)
   + $6,52 0 (No. of sea days)

Estimation of Shipping Cost 83

$\quad\quad\quad$ + $440 (No. of port days)
$\quad\quad\quad$ + $1,300] / 33,250 tons

10. **40,000 DWT, U.S.-flag ship**

$\quad\quad$ T.C./ton = 1.4 [$45,331 (No. of voyage days)
$\quad\quad\quad$ + $6,524 (No. of sea days)
$\quad\quad\quad$ + $490 (No. of port days)
$\quad\quad\quad$ + $1,520] / 38,000 tons

11. **40,000 DWT, Japan-flag ship**

$\quad\quad$ T.C./ton = 1.4 [$20,356 (No. of voyage days)
$\quad\quad\quad$ + $6,524 (No. of sea days)
$\quad\quad\quad$ + $490 (No. of port days)
$\quad\quad\quad$ + $1,520] / 38,000 tons

12. **40,000 DWT, Liberia-flag ship**

$\quad\quad$ T.C./ton = 1.4 [$16,456 (No. of voyage days)
$\quad\quad\quad$ + $6,524 (No. of sea days)
$\quad\quad\quad$ + $490 (No. of port days)
$\quad\quad\quad$ + $1,520] / 38,000 tons

$\quad\quad$ The above equations are used to calculate shipping rate for each type of ship and to set up transportation rate matrices in order to construct transportation models to find the optimal shipping patterns for the United States southern region rice exports. Therefore, they are rewritten as follows to make the variables the number of days at sea and number of days in port.

1. **25,000 DWT, U.S.-flag ship**

$\quad\quad$ T.C./ton = 1.4 [days in port ($32,877 + $350)
$\quad\quad\quad$ + days at sea ($32,877 + $6,524)
$\quad\quad\quad$ + $1,030] / 23,750 tons
$\quad\quad\quad$ = 1.4 [$33,227 (days in port)
$\quad\quad\quad\quad$ + $39,401 (days at sea) + $1,030] / 23,750tons
$\quad\quad\quad$ = 1.959 (days in port) + 2.323 (days at sea)+0.061

2. *25,000 DWT, Japan.-flag ship*

   T.C./ton = 1.4 [days in port ($15,657 + $350)
   + days at sea ($15,657 + $6,524)
   + $1,030] / 23,750 tons
   = 0.934 (days in port) + 1.307 (days at sea) + 0.061

3. *25,000 DWT, Liberia-flag ship*

   T.C./ton = 1.4 [days in port ($11,757 + $350)
   + days at sea (11,757 + $6,524)
   + $1,030] / 23,750 tons
   = 0.714 (days in port + 1.078 (days at sea) + 0.061

4. *30,000 DWT, U.S.-flag ship*

   T.C./ton = 1.4 [days in port ($37,034 + $380)
   + days at sea ($37,034 + $6,524)
   + $1,245] / 28,500 tons
   = 1.838 (days in port) + 2.14 (days at sea) + 0.061

5. *30,000 DWT, Japan-flag ship*

   T.C./ton = 1.4 [days in port ($17,224 + $380)
   + days at sea ($17,224 + $6,524)
   + $1,245] / 28,500 tons
   = 0.865 (days in port) + 1.167 (days at sea) + 0.061

6. *30,000 DWT, Liberia-flag ship*

   T.C./ton = 1.4 [days in port ($13,324 + $380)
   + days at sea ($13,324 + $6,524)
   + $1,245] / 28,500 tons
   = 0.673 (days in port) + 0.975 (days at sea) + 0.061

7. *35,000 DWT, U.S.-flag ship*

   T.C./ton = 1.4 [days in port ($41,182 + $440)
   + days at sea ($41,182 + $6,524)

+ $1,300] / 33,250 tons
= 1.753 (days in port) + 2.009 (days at sea) + 0.055

8. *35,000 DWT, Japan-flag ship*

T.C./ton = 1.4 [days in port ($18,790 + $440)
+ days at sea ($18,790 + $6,524)
+ $1,300] / 33,250 tons
= 0.810 (days in port) + 1.066 (days at sea) + 0.055

9. *35,000 DWT, Liberia-flag ship*

T.C./ton = 1.4 [days in port ($14,890 + $440)
+ days at sea ($14,890 + $6,524)
+ $1,300] / 33,250 tons
= 0.646 (days in port) + 0.902 (days at sea) + 0.055

10. *40,000 DWT, U.S.-flag vessel*

T.C./ton = 1.4 [days in port ($45,331 + $490)
+ days at sea ($45,331 + $6,524)
+ $1,520] / 38,000 tons
= 1.688 (days in port) + 1.910 (days at sea) + 0.056

11. *40,000 DWT, Japan-flag vessel*

T.C./ton = 1.4 [days in port ($20,356 + $490)
+ days at sea ($20,356 + $6,524)
+ $1,520] / 38,000 tons
= 0.768 (days in port) + 0.991 (days at sea) + 0.056

12. *40,000 DWT, Liberia-flag vessel*

T.C./ton = 1.4 [days in port ($16,456 + $490)
+ days at sea ($16,456 + $6,524)
+ $1,520] / 38,000 tons
= 0.625 (days in port) + 0.847 (days at sea) + 0.056

Unit transportation rates per ton for each of the three flags, and four sizes of ships over each possible shipping route are calculated from the above twelve relationships and the actual at-sea days and in-port days, and the results are listed in tables 22 through 25. Those shipping rates are then used to construct transportation models in Chapter IV.

SUMMARY

Transportation costs are one of the most important factors in determining the routes over which U.S. rice is shipped. Since published transportation cost data are not available for all possible international routes of rice shipment, estimation of shipping costs becomes the first necessary step before transportation models are established.

There are two approaches in estimating shipping cost: the statistical approach and the engineering approach. In this book an economic-engineering approach is used to measure cost relationships. In order to get necessary variables used in cost estimation, such as distance between each possible pair of ports and days a ship spends during one voyage, including days in port and days at sea, seven ports of origin and thirteen ports of destination are selected as base points representing certain U.S. southern exporting regions and certain world importing areas, through which all shipments can be made.

The number of days in ports and the number of days spent at sea combined are the number of voyage days a ship needs to complete a shipping voyage. This is also an important factor in determining shipping rates. The number of days spent in ports is obtained by contacting the major U.S. southern port authorities, and the number of at-sea days is calculated by dividing the distance between each pair of ports by the number of miles each ship travels during one day.

Several assumptions for estimating shipping costs are stated. In this book, four ship sizes and three types of ship flags are used in order to compare the differences in shipping costs. There are three broad components in transportation costs: vessel ownership expenses, at-sea expenses and in-port expenses. Shipping expenses in each group are estimated. These values are then combined to serve as the basis to calculate the unit transportation rates per ton over each possible shipping route for vessels of different sizes and different flags.

Table 22 Calculated Transportation Rate for Rice Exports
(For *25,000 DWT vessels*)

Origin Destination	(1)	(2)	(3)	(4) *(dollars / ton)*	(5)	(6)	(7)
1. (CAC) Kingston (Jamaica)							
U.S.-flag	14.9	15.8	14.9	21.7	13.6	17.8	15.8
Japan-flag	7.8	8.3	7.8	11.1	7.2	9.2	8.3
Liberia-flag	6.2	6.6	6.2	8.7	5.8	7.3	6.6
2. (WSA) Callao (Peru)							
U.S.-flag	27.0	27.9	27.0	34.7	13.9	29.9	28.1
Japan-flag	14.6	15.1	14.6	18.4	14.0	16.0	15.2
Liberia-flag	11.8	12.2	11.8	14.8	11.4	12.9	12.3
3. (ESA) Rio de Janeiro (Brazil)							
U.S.-flag	38.8	40.0	39.0	47.2	38.2	42.4	40.7
Japan-flag	21.2	21.9	21.4	25.5	21.1	23.1	22.3
Liberia-flag	17.3	17.8	17.4	20.3	17.2	18.7	18.1
4. (NCE) Notterdam (Netherlands)							
U.S.-flag	37.2	38.3	37.4	45.6	36.6	41.0	39.0
Japan-flag	20.3	21.0	20.4	24.6	20.1	22.3	21.4
Liberia-flag	16.5	17.0	16.6	19.8	16.4	18.1	17.4

(continued on next page)

	(1)	(2)	(3)	(4)	(5)	(6)	(7)
5. (NSE) Valencia (Spain)							
U.S.-flag	37.9	39.0	38.3	46.3	37.5	41.7	40.0
Japan-flag	20.7	21.4	20.4	25.0	20.7	22.7	21.9
Liberia-flag	16.8	17.4	17.0	20.2	16.9	18.4	17.8
6. (WA) Aqaba (Jordan)							
U.S.-flag	51.3	52.7	52.0	59.8	50.8	50.2	53.3
Japan-flag	28.3	29.1	28.7	32.6	28.1	30.3	29.3
Liberia-flag	23.1	23.7	23.4	26.4	23.0	24.7	23.9
7. (SSA) Singapore (Singapore)							
U.S.-flag	79.9	80.8	79.9	88.1	79.1	83.3	81.5
Japan-flag	44.4	44.9	44.4	48.5	44.1	46.1	45.3
Liberia-flag	36.3	36.8	36.3	39.6	36.2	37.7	37.1
8. (EA) Hong Kong							
U.S.-flag	74.3	75.5	74.6	82.3	77.3	77.5	75.7
Japan-flag	41.2	41.9	41.3	45.2	40.8	42.8	42.0
Liberia-flag	33.8	34.3	33.8	37.3	33.5	35.0	34.4
9. (AO) Wellington (New Zealand)							
U.S.-flag	58.3	59.5	58.3	66.0	57.1	61.2	59.5
Japan-flag	32.2	32.9	32.2	36.1	31.7	33.7	32.9
Liberia-flag	26.3	26.8	26.3	29.8	25.9	27.5	26.9

(*continued on next page*)

	(1)	(2)	(3)	(4)	(5)	(6)	(7)
10. (WAF) Monrovia (Liberia)							
U.S.-flag	37.2	38.3	37.4	45.6	36.6	40.8	39.0
Japan-flag	20.3	21.0	20.4	24.6	20.1	22.2	21.4
Liberia-flag	16.5	17.0	16.1	19.8	16.4	18.0	17.4
11. (NAF) Algiers (Algeria)							
U.S.-flag	38.1	39.3	38.3	46.5	37.5	41.7	40.0
Japan-flag	20.8	21.5	21.0	25.1	20.7	22.7	21.9
Liberia-flag	16.9	17.5	17.0	20.3	16.9	18.4	17.8
12. (EAF) Maputo (Mozambique)							
U.S.-flag	65.5	66.4	65.5	73.2	64.3	68.6	66.7
Japan-flag	36.3	36.8	36.3	40.1	35.7	37.9	36.9
Liberia-flag	29.7	30.1	29.7	32.7	29.3	30.9	30.2
13. (SAF) Cape Town (S. Africa)							
U.S.-flag	51.8	53.0	52.0	60.2	51.2	55.4	53.7
Japan-flag	28.5	29.2	28.7	32.8	28.4	30.4	29.6
Liberia-flag	23.3	23.8	23.4	26.6	23.2	24.7	24.2

Note: Column headings are as follows: (1) Mobile; (2) Baton Rouge; (3) New Orleans; (4) Beaumont; (5) Galveston; (6) Houston; and (7) Port Arthur.

Table 23 Calculated Transportation Rate for Rice Exports
(For 30,000 DWT vessels)

Origin Destination	(1)	(2)	(3)	(4) (dollars / ton)	(5)	(6)	(7)
1. (CAC) Kingston (Jamaica)							
U.S.-flag	13.8	14.7	13.8	22.0	12.6	16.5	14.7
Japan-flag	7.0	7.5	7.0	11.0	6.5	8.4	7.5
Liberia-flag	5.7	6.1	5.7	8.8	5.3	6.7	6.1
2. (WSA) Callao (Peru)							
U.S.-flag	25.0	25.8	25.0	34.0	23.8	27.7	26.0
Japan-flag	13.1	13.6	13.1	17.5	12.6	14.4	13.7
Liberia-flag	10.8	11.1	10.8	14.2	10.4	11.8	11.2
3. (ESA) Rio de Janeiro (Brazil)							
U.S.-flag	35.9	37.0	36.1	45.6	35.3	39.2	39.4
Japan-flag	19.1	19.6	19.2	23.8	18.9	20.7	20.0
Liberia-flag	15.7	16.2	15.8	19.5	15.6	17.1	16.5
4. (NCE) Notterdam (Netherlands)							
U.S.-flag	34.4	35.5	34.6	44.1	33.8	37.9	36.1
Japan-flag	18.2	18.8	18.4	23.0	18.1	20.0	19.2
Liberia-flag	15.0	15.5	15.1	18.0	15.0	16.5	15.8

(continued on next page)

	(1)	(2)	(3)	(4)	(5)	(6)	(7)
5. (NSE) Valencia (Spain)							
U.S.-flag	35.0	36.1	35.5	44.7	34.7	38.6	37.0
Japan-flag	18.6	19.2	18.8	23.3	18.5	20.4	19.6
Liberia-flag	15.3	15.8	15.5	19.1	15.3	16.8	16.2
6. (WA) Aqaba (Jordan)							
U.S.-flag	47.4	48.7	48.1	57.1	46.9	51.0	49.1
Japan-flag	25.4	26.1	25.7	30.1	25.2	27.1	26.3
Liberia-flag	21.0	21.6	21.3	24.8	20.9	22.4	21.8
7. (SSA) Singapore (Singapore)							
U.S.-flag	73.8	74.6	73.8	83.2	70.0	76.9	75.3
Japan-flag	39.7	40.2	39.7	44.3	39.4	41.3	40.5
Liberia-flag	33.0	33.4	33.0	36.7	32.8	34.2	33.7
8. (EA) Hong Kong							
U.S.-flag	68.6	69.7	68.8	77.9	67.6	71.5	69.9
Japan-flag	36.9	37.5	37.0	41.4	36.5	38.4	37.6
Liberia-flag	30.6	31.1	30.7	34.2	30.4	31.8	31.2
9. (AO) Wellington (New Zealand)							
U.S.-flag	53.9	54.9	53.9	62.9	52.7	56.6	54.9
Japan-flag	28.9	29.4	28.9	33.2	28.3	30.1	29.4
Liberia-flag	23.9	24.4	23.9	27.4	23.5	25.0	24.4

(*continued on next page*)

	(1)	(2)	(3)	(4)	(5)	(6)	(7)
10. (WAF) Monrovia (Liberia)							
U.S.-flag	34.4	35.5	34.6	44.1	33.8	37.7	36.1
Japan-flag	18.2	18.1	18.4	23.0	18.1	19.9	19.2
Liberia-flag	15.0	15.5	15.1	18.8	15.0	16.4	15.8
11. (NAF) Algiers (Algeria)							
U.S.-flag	35.2	36.3	35.5	44.9	34.7	38.6	37.0
Japan-flag	18.7	19.3	18.8	23.4	18.5	20.4	19.6
Liberia-flag	15.4	15.9	15.5	19.2	15.3	16.8	16.2
12. (EAF) Maputo (Mozambique)							
U.S.-flag	60.5	61.3	60.5	69.6	59.3	60.4	61.6
Japan-flag	32.5	32.9	32.5	36.9	32.0	33.9	33.1
Liberia-flag	26.9	27.3	26.9	30.4	26.6	28.1	27.4
13. (SAF) Cape Town (S. Africa)							
U.S.-flag	47.9	48.9	48.1	57.6	47.3	51.2	49.6
Japan-flag	25.6	26.2	25.7	30.3	25.4	27.3	26.5
Liberia-flag	21.2	21.7	21.3	25.0	21.1	22.5	22.0

Note: Column headings are as follows: (1) Mobile; (2) Baton Rouge; (3) New Orleans; (4) Beaumont; (5) Galveston; (6) Houston; and (7) Port Arthur.

Table 24 Calculated Transportation Rate for Rice Exports
         (For 35,000 DWT vessels)

Origin Destination	(1)	(2)	(3)	(4) (*dollars / ton*)	(5)	(6)	(7)
1. (CAC) Kingston (Jamaica)							
U.S.-flag	14.4	15.2	14.4	20.4	13.2	16.9	15.2
Japan-flag	7.1	7.3	7.1	10.0	6.2	8.4	7.5
Liberia-flag	5.9	6.2	5.9	8.2	5.5	6.9	6.2
2. (WSA) Callao (Peru)							
U.S.-flag	24.1	25.6	24.8	31.7	23.7	27.4	25.8
Japan-flag	12.0	13.1	12.7	15.9	12.2	13.9	13.2
Liberia-flag	10.2	10.9	10.5	13.2	10.2	11.6	11.0
3. (ESA) Rio de Janeiro (Brazil)							
U.S.-flag	33.8	36.1	35.3	42.3	34.5	38.2	36.7
Japan-flag	17.1	18.6	18.2	21.7	17.9	19.7	19.0
Liberia-flag	14.5	15.6	15.2	17.8	15.0	16.4	15.9
4. (NCE) Notterdam (Netherlands)							
U.S.-flag	32.5	34.7	33.9	40.1	33.1	37.0	35.3
Japan-flag	16.4	17.9	17.5	20.9	17.2	19.0	18.2
Liberia-flag	13.9	15.0	14.6	17.4	14.4	15.9	15.2

(*continued on next page*)

Origin	(1)	(2)	(3)	(4)	(5)	(6)	(7)
5. (NSE)							
Valencia							
(Spain)							
U.S.-flag	33.1	35.3	34.7	41.7	33.9	37.6	36.1
Japan-flag	16.7	18.2	17.9	21.3	17.6	19.3	18.6
Liberia-flag	14.1	15.2	15.0	17.7	14.8	16.2	15.6
6. (WA)							
Aqaba							
(Jordan)							
U.S.-flag	43.1	47.1	46.5	53.4	45.4	49.3	47.5
Japan-flag	22.4	24.5	24.5	27.5	23.7	25.5	24.7
Liberia-flag	19.1	20.6	20.3	22.9	19.9	21.4	20.7
7. (SSA)							
Singapore							
(Singapore)							
U.S.-flag	67.6	71.4	70.6	77.7	69.9	73.6	72.0
Japan-flag	34.6	37.4	37.0	40.3	36.7	38.4	37.7
Liberia-flag	28.5	31.5	31.1	33.9	30.9	32.3	31.7
8. (EA)							
Hong Kong							
U.S.-flag	65.9	66.8	66.0	72.9	64.9	68.6	67.0
Japan-flag	33.2	34.9	34.5	37.8	34.0	35.8	35.0
Liberia-flag	28.3	29.4	29.0	29.1	28.7	30.0	29.5
9. (AO)							
Wellington							
(New Zealand)							
U.S.-flag	49.9	53.0	51.9	58.8	50.8	54.5	53.0
Japan-flag	25.4	27.6	27.0	30.3	26.6	28.3	27.6
Liberia-flag	21.6	23.2	22.7	25.4	22.3	23.7	23.2

(continued on next page)

Estimation of Shipping Cost

Origin	(1)	(2)	(3)	(4)	(5)	(6)	(7)
10. (WAF) Monrovia (Liberia)							
U.S.-flag	32.5	34.7	33.9	40.9	33.1	36.8	35.3
Japan-flag	16.4	17.9	17.5	20.8	17.2	18.9	18.2
Liberia-flag	13.9	15.0	14.6	17.3	14.4	15.8	15.2
11. (NAF) Algiers (Algeria)							
U.S.-flag	33.3	35.5	34.7	41.7	33.9	37.6	36.1
Japan-flag	16.8	18.2	17.9	21.3	17.6	19.3	18.6
liberia-flag	14.2	15.4	15.0	17.7	14.8	16.2	15.6
12. (EAF) Maputo (Mozambique)							
U.S.-flag	55.8	59.0	58.2	64.8	57.0	60.9	59.2
Japan-flag	28.5	30.8	30.4	33.5	29.9	31.7	30.9
Liberia-flag	24.2	25.9	25.5	28.1	25.1	26.6	26.0
13. (SAF) Cape Town (S. Africa)							
U.S.-flag	34.5	47.3	46.5	53.6	45.8	49.5	47.9
Japan-flag	22.6	24.6	24.2	27.5	23.9	25.6	24.9
Liberia-flag	19.2	20.6	20.3	22.9	20.1	21.5	20.9

Note: Column headings are as follows: (1) Mobile; (2) Baton Rouge; (3) New Orleans; (4) Beaumont; (5) Galveston; (6) Houston; and (7) Port Arthur.

Table 25 Calculated Transportation Rate for Rice Exports
(For 40,000 DWT vessels)

Origin Destination	(1)	(2)	(3)	(4)	(5)	(6)	(7)
				(dollars / ton)			
1. (CAC) Kingston (Jamaica)							
U.S.-flag	14.3	14.9	14.2	20.0	13.0	16.6	14.9
Japan-flag	7.0	7.3	6.9	9.6	6.4	8.1	7.3
Liberia-flag	5.8	6.1	5.8	8.0	5.4	6.7	6.1
2. (WSA) Callao (Peru)							
U.S.-flag	24.1	24.8	24.1	30.7	23.0	26.5	25.1
Japan-flag	12.0	12.4	12.0	15.1	11.6	13.2	12.5
Liberia-flag	10.2	10.5	10.2	12.7	9.8	11.1	10.6
3. (ESA) Rio de Janeiro (Brazil)							
U.S.-flag	33.8	34.8	34.0	41.0	33.3	36.9	35.4
Japan-flag	17.1	17.6	17.2	20.5	16.9	18.6	17.9
Liberia-flag	14.5	14.9	14.6	17.3	14.4	15.7	15.2
4. (NCE) Notterdam (Netherlands)							
U.S.-flag	32.5	33.4	32.7	39.6	31.9	35.7	34.0
Japan-flag	16.4	16.9	16.5	19.8	16.2	18.0	17.2
Liberia-flag	13.9	14.3	14.0	16.7	13.8	15.2	14.6

(continued on next page)

Estimation of Shipping Cost

Origin	(1)	(2)	(3)	(4)	(5)	(6)	(7)
5. (NSE) Valencia (Spain)							
U.S.-flag	33.1	34.0	33.6	40.2	32.7	36.3	34.8
Japan-flag	16.7	17.2	16.9	20.1	16.6	18.3	17.6
Liberia-flag	14.1	14.6	14.3	16.9	14.1	15.4	14.9
6. (WA) Aqaba (Jordan)							
U.S.-flag	44.1	45.3	44.7	51.3	43.6	47.4	45.7
Japan-flag	22.4	23.0	22.7	25.8	22.3	24.0	23.2
Liberia-flag	19.1	19.6	19.3	21.8	18.9	20.4	19.7
7. (SSA) Singapore (Singapore)							
U.S.-flag	67.6	68.4	67.6	74.4	66.9	70.5	69.0
Japan-flag	34.6	35.0	34.6	37.8	34.4	36.0	35.3
Liberia-flag	29.5	29.8	29.5	30.1	29.3	30.6	30.1
8. (EA) Hong Kong							
U.S.-flag	65.9	64.0	63.2	69.8	62.1	65.7	64.2
Japan-flag	33.2	32.8	32.4	35.4	31.9	33.5	32.9
Liberia-flag	28.3	27.9	27.5	29.1	27.2	28.5	27.9
9. (AO) Wellington (New Zealand)							
U.S.-flag	49.9	50.8	49.9	56.4	48.8	52.3	50.8
Japan-flag	25.4	25.9	25.4	28.5	24.9	26.6	25.9
Liberia-flag	21.6	22.0	21.6	24.1	21.2	22.6	22.0

(*continued on next page*)

Origin	(1)	(2)	(3)	(4)	(5)	(6)	(7)
10. (WAF) Monrovia (Liberia)							
U.S.-flag	32.5	33.4	32.7	39.3	32.0	35.5	34.0
Japan-flag	16.4	16.9	16.5	19.6	16.2	17.9	17.2
Liberia-flag	13.9	14.3	14.0	16.5	13.8	15.1	14.6
11. (NAF) Algiers (Algeria)							
U.S.-flag	33.3	34.2	33.4	40.2	32.7	36.3	34.8
Japan-flag	16.8	17.3	16.9	20.1	16.6	18.3	17.6
Liberia-flag	14.2	14.6	14.3	16.9	14.1	15.4	14.9
12. (EAF) Maputo (Mozambique)							
U.S.-flag	55.8	56.6	55.8	62.2	54.7	58.4	56.7
Japan-flag	28.5	28.9	28.5	31.5	28.0	29.8	29.0
Liberia-flag	24.2	24.6	24.2	26.7	23.8	25.3	24.6
13. (SAF) Cape Town (S. Africa)							
U.S.-flag	44.5	45.5	44.7	51.8	44.0	47.5	46.1
Japan-flag	22.6	23.1	22.7	25.9	22.5	24.1	23.4
Liberia-flag	19.2	19.6	19.3	21.9	19.1	20.4	19.9

Note: Column headings are as follows: (1) Mobile; (2) Baton Rouge; (3) New Orleans; (4) Beaumont; (5) Galveston; (6) Houston; and (7) Port Arthur.

NOTES

1. S. G. Sturney. *Shipping Economics*. London: MacMillan Press, Ltd., 1975.
2. M. LeRoy Davis. *Cost of Shipping United States Grain Exports to Principal World Markets*. Unpublished Master Thesis, Department of Agricultural Economics, Iowa State University. 1968
3. A. C. Winter. Maritime Administration. US. Department of Commerce, New York. *Data on Shipping Costs.* 1968.
4. John William Uhrig. *Economic Effects of Charges on Transportation Rates and Processing Capacity of Soybean Procurement by Iowa processors.* Unpublished Ph.D. dissertation. Iowa State University, Ames, Iowa. 1965.
5. U.S. Department of Commerce, Bureau of the Census. *Unpublished Foreign Trade Statistics*, SA705/705IT, 1986. Bureau of the Census, U.S. Department of Commerce, Washington, D.D., 1986.
6. B. John Harrer. *Ocean Freight Rates and Agricultural Trade.* Unpublished thesis, Purdue University. June 1979.
7. Suk-Won Yoon. *A Special Equilibrium Analysis of the Competitive Position of the Southern U S. Rice Industry in the international Market.* Unpublished Ph.D. dissertation. Mississippi State University. May 1988.
8. Reed's Marine Distance Tables. Published by Thomas Reed Publications Limited, London. 1978.
9. One knot is one nautical mile per hour.
10. A nautical mile is one minute of a great circle of the earth equal to 6,080 feet.
11. Roy L. Nersesian. *Ships and Shipping*. Tulsa, Oklahoma. 1981.
12. James E. Caponiti., Chief, Division of Ship Operating Costs, Maritime Administration, U.S. Department of Transportation. Private Communication. October 1987, December 1988.
13. A. Charnes and W.W. Cooper. The Stepping Stone Method of Explaining Linear Programming Calculations in Transportation Problems. Management Science, 1:49-69.

14. DWT: Deadweight tonnage measures the total carrying capacity of the vessel (cargo plus fuel, fresh water, store, etc.). It is the unit that would normally be used for assessing the impact of the world shipbuilding industry on the shipping market supply/demand balance.

15. Maritime Research Incorporated. Weekly published data on ship charters, New York, 1972-1986.

16. National Academy of Sciences-National Research Council. *Maritime Transportation of Unitized Cargo: A Comparative Economic Analysis of Break Bulk and Unit Load Systems*. National Academy of Sciences-National Research Council Publication 745. 1959.

17. T.C.C. refers to total construction cost. The estimating procedures for vessel construction cost are shown in Appendix A.

18. The 1.4 was calculated from the assumption of 60 percent of a normal full load on the return trip.

IV

The Transportation Models

In order to test the transportation rates developed in Chapter III, transportation models are established to determine a least-cost shipping pattern on the basis of the estimated shipping rates. The selection of a model for any research should contribute to the objectives of the study. On the other hand, complex problems require qualification into workable, conceptually feasible computational models. Therefore, cost minimization is selected as the primary goal of this study and transportation models are employed for the analysis where the objective is to transport a single homogenous commodity—in this book, rice—from a number of U.S. southern ports of origin to a number of international ports of destinations at a minimum total cost.

METHODOLOGY AND DATA REQUIREMENTS

Transportation Model

The term "transportation problem" is used to refer to a subclass of a linear programming problem for which computational procedures have been developed which take advantage of the special structure of the model. It has been used to identify the optimal flow patterns of shipping rice from the U.S. southern region to world importing countries, in which fixed supplies in each surplus region are to be allocated to meet fixed demands in each deficit region so as to minimize total transfer costs. By use of the transportation model of linear programming, the economists are able to make great strides toward quantifying the spatial advantages of different regions.

The transportation problem was first developed by Hitchcock and Koopmans.[1] Hitchcock applied the procedure to the problem of minimizing the cost of distributing a product from several factories to a number of cities. He used a geometric approach to solve the problem. Dantzing formulated the transportation problem within the framework of linear programming and then developed a special form of the simplex technique for solving the problem.[2] Charnes and Cooper developed a procedure suggested by Dantzing for obtaining an initial feasible solution to the transportation problem which was called the "northwest-corner rule."[3]

Other methods for solving the transportation problem have since been developed. In this book, a linear programming formulation of the transportation model is utilized by employing *SPERRY UNIVAC's Functional Mathematical Programming System (FMPS)*. For analytical purposes, six models are developed to determine optimal shipping patterns of rice trade from U.S. southern exporting regions to world importing regions by minimizing the total transportation costs.

The first model (*Basic Model*) uses the transportation costs estimated in Chapter III. The other five models are constructed with different transportation costs. Together, the six models are used to compare the effects of differences in shipping costs on the shipping patterns. With the first model being the basic one, the second model assumes that for the optimal size ships determined in the basic model, interest rates in estimating the shipping cost are increased by 20 percent. In the third model, for the optimal size vessels selected by the basic model, the insurance rates are increased by 20 percent so that the estimated shipping rates are changed.

For the fourth model, it is assumed that the port expenses in ports of New Orleans and Houston are increased by 20 percent which in turn affect the estimated transportation costs from these ports. The fifth model assumes that the operating costs for U.S.-flag vessels are reduced by 50 percent and in the last model, the construction costs for U.S.-flag ships are reduced by 50 percent. The transportation costs estimated from these changes are found in Appendix B tables 1 through 9. These different models provide an opportunity to analyze the effects of these factors on the least-cost shipping patterns of U.S. rice exports separately and collectively.

Mathematical Structure

Suppose there are *m* exporting regions (m = 7 in this model) supplying *n* importing regions (n = 13 in this model). The exporting regions supply rice at levels $S_1, S_2, \ldots, S_m$ and the quantity demanded at the importing regions for rice is $D_1, D_2, \ldots, D_n$. The problem now becomes one of determining the shipping pattern which minimizes the total transportation costs.

If the unit cost of shipping rice from exporting region i to importing region j is represented by T_{ij}, and X_{ij} represents the amount of rice shipped from exporting region i to importing region j, the mathematical notation of the problem can be expressed in the following form:[4]

$$\text{Minimize TC} = \sum_{i=1}^{m} \sum_{j=1}^{n} X_{ij} T_{ij}, \quad i = 1, 2, \ldots, n \quad j = 1, 2, \ldots, m$$

Subject to:

(1) $\sum_{i=1}^{m} X_{ij} = D_j$

—The sum of what arrives at each importing region from various origins is equal to the demand at that importing area.

(2) $\sum_{j=1}^{n} X_{ij} = S_i$

—The sum of what leaves each origin for various importing regions is equal to the supplies of that surplus area.

(3) $\sum_{i=1}^{m} S_i = \sum_{j=1}^{n} D_j$

—Total supplies equal total demand

(4) $X_{ij} \geq 0$ for all i, j
$T_{ij} \geq 0$ for all i, j
$S_i \geq 0$ for all i
$D_j \geq 0$ for all j

—This means that negative shipments, transfer cost, supply and demand have no physical meaning.

Data Requirement

As mentioned earlier, the analysis of rice shipments is based on data obtained from unpublished Foreign Trade Statistics Reports, SA705, 705IT.[5] The following information for each shipment is collected: (1) United States southern port region, (2) United States southern ports of origin, (3) countries and regions of destination, (4) commodity, (5) number of pounds and tons of the commodity shipments, (6) dollar value, and (7) ship type.

Data are collected for the period 1981-1986, and the 1986 calendar year is chosen as the data base for this study. While a number of reasons could be given for selecting this base, the overriding one deals with the availability and applicability of the data. While industry parameters may change from year to year, basic changes in structure of the industry occur rather slowly and over an extended period of time.[6] Therefore, in selecting the base year, availability and quality of data are deemed to be more important than timeliness of data.

Given this consideration, the 1986 calendar year is the base year for the analysis. That is to say, the amount available for export from each port of origin is estimated from 1986 United States southern rice exports. Likewise the distribution of exports among deficit areas of the world is based on the 1986 experience. This should test the cost data estimated in the previous chapter and give the total outlay for transportation of United States southern rice into the world markets if it is transported at least cost and under the specific assumptions of the model.

Therefore, the amount (tons) of rice available for export from each of the seven points of surplus region and the amount of rice deficit in the 13 points of market areas are based on the actual amounts shipped from and received at these regions during 1986. The rice surplus for export by ports of origin for the period of 1981-1986 is listed in table 26. The deficit of rice by market area for the period 1981-1986 is given in table 27. Actual shipping pattern for rice exports from the seven U.S. southern ports to the thirteen importing regions in the world in 1986 is calculated and presented in table 28.

Transportation costs are an integral part of the data required to solve the transportation problem. Transportation rates for shipping rice in terms of U.S. dollars per ton have been estimated from the equations developed in Chapter III and listed in tables 22 through 25 for the four sizes (25,000 DWT, 30,000 DWT, 35,000 DWT and 40,000 DWT) and three flags (U.S.-flag, Japan-flag and Liberia-flag) vessels.

Table 26 Rice Surplus Exported from U.S. Southern Region *(short tons)*

Region	1981	1982	1983
Mobile	152,838.3	115,530.6	20,582.3
Baton Rouge	149,642.2	33,982.8	21,467.3
New Orleans	254,089.7	419,238.4	479,781.4
Beaumont	13,340.8	22,291.6	7,287.8
Galveston	423,521.5	320,399.8	350,814.9
Houston	720,020.8	740,756.0	407,281.5
Port Arthur	639,712.8	403,217.3	354,061.1
Total	2,353,166.2	2,055,416.5	1,641,276.3

Region	1984	1985	1986
Mobile	9,570.7	4,199.1	764.2
Baton Rouge	---	59,258.0	64,773.2
New Orleans	191,096.0	317,790.3	260,623.5
Beaumont	---	11.0	146.3
Galveston	287,490.3	202,542.8	234,065.4
Houston	527,184.9	362,478.4	752,755.9
Port Arthur	166,634.8	338,270.5	901,284.3
Total	1,181,976.7	1,284,550.2	2,214,412.8

Source: U.S. Department of Commerce, Bureau of the census, *Unpublished Foreign Trade Statistics,* 1981-1986. Bureau of the Census, U.S. Department of Commerce, Washington, D.C., 1987

Table 27 Rice Deficit by World Region (*in short tons*)

Region	1981	1982	1983
Central America & Caribbean	81,517.1	25,954.9	89,645.7
Western Southern America	93,120.1	3,304.5	117,598.8
Eastern Southern America	---	225.6	388.1
Northwestern & Central Europe	330,344.3	389,059.6	367,375.3
Northeastern & Southern Europe	89,764.7	100,833.6	14,550.9
Eastern Asia	296,553.1	1,183.5	---
Western Asia	525,098.5	636,543.7	453,343.2
Southern & Southeastern Asia	120,293.7	44,280.7	41,012.4
Australia & Oceania	6,319.7	1,876.9	1,073.1
Western Africa	634,502.9	584,333.9	276,321.4
Northern Africa	8,304.7	5,815.9	6,681.2
Eastern Africa	43,151.8	125,039.0	114,180.9
Southern Africa	124,195.6	136,964.7	159,105.3
Total	2,353,166.2	2,055,416.5	1,641,276.3

Region	1984	1985	1986
Central America & Caribbean	36,608.7	99,758.3	182,006.8
Western Southern America	48,081.2	4,691.1	8,595.6
Eastern Southern America	268.3	883.3	607,784.8
Northwestern & Central Europe	255,861.0	245,409.6	323,264.4
Northeastern & Southern Europe	7,617.2	57,819.5	22,042.9
Eastern Asia	100.7	---	727.5
Western Asia	555,334.5	489,711.0	637,960.9
Southern & Southeastern Asia	28,701.8	44,747.0	3,457.7
Australia & Oceania	1,192.7	629.8	2,561.5
Western Africa	160,909.3	204,975.3	253,999.4
Northern Africa	213.7	12,956.6	311.5
Eastern Africa	51,620.0	63,659.8	85,378.4
Southern Africa	35,467.6	59,308.9	86,321.4
Total	1,181,976.7	1,284,550.2	2,214,412.8

Source: U.S. Department of Commerce, Bureau of the Census. Unpublished Foreign Trade Statistics, 1981-1986. Bureau of the Census, U.S. Department of Commerce, Washington, D.C., 1987.

ASSUMPTIONS OF THE MODEL

Since the transportation model is a subclass of linear programming, it involves certain restrictive assumptions that are necessary for the construction and application of the model. They are outlined as follows:[7]

1. Even though there are many different kinds of rice, for the purpose of analysis, all rice resources and products are assumed to be homogenous. This means that the supply of product at any region or origin serves equally well to satisfy the demands of any consuming sector.
2. The supplies of products available at the various origins and the demands of the various destinations are known and fixed. That is, excess supply and excess demand are readily available in each exporting and importing region, also total demand equals total supply.
3. Transportation rates are known and constant per unit. In other words, the costs of moving the commodity from origins to destinations are known and are independent of the number of units moved.
4. The objective of the models is to minimize the total costs.
5. Non-importing countries and non-exporting regions do not have any influence on international trade.
6. All shipments in the models are assumed to be made on vessels of four different sizes: 25,000 DWT, 30,000 DWT, 35,000 DWT, and 40,000 DWT; and of three different flags: U.S.-flag, Japan-flag, and Liberia-flag.

PROCEDURE OF THE MODEL

Since the overall objective of this study is to find and examine the least-cost shipping patterns for the United States southern rice exports, certain transportation models have been employed for the analysis which are concerned with the cost minimizing allocation of known supplies in the U.S. southern surplus markets or regions to the world deficit markets or regions within the market area.

108   *Competitiveness in U.S. Grain Exports*

The relevant data, regional supply and regional demand have been mentioned in previous chapters and are presented in tables 11, 12, 26 and 27. Thirteen deficit region groups, seven surplus region groups, and the respective base points have been determined for the study. They are listed in tables 13 through 15. In Chapter III, transportation rates needed between each possible pair of locations for each type of ship have been estimated and are shown in tables 22 through 24. Optimal shipping patterns and the total cost of transportation are determined for the six models by using the FMPS technique with the given data.

ANALYSIS OF THE RESULTS

Optimizing the models described in this chapter provides two types of output that are of interest: (1) the optimum shipping pattern for the rice shipments with vessels of all four sizes and three flags; and (2) total cost of transportation for all the six optimal shipping patterns. All the results of the models, of course, are dependent on the transportation rates estimated in Chapter III and the assumptions for each model.

The total costs of transportation for the six models are as follows:

Model   I: T.C. = $42,340,273.50
Model   II: T.C. = $43,666,216.50
Model   III: T.C. = $42,571,701.50
Model   IV: T.C. = $42,340,460.50
Model   V: T.C. = $42,340,273.50
Model   VI: T.C. = $42,340,273.50

The optimum shipping patterns for the six models are presented in tables 29 through 34.

Transportation Cost

All the results of the models depend on the estimated transportation costs. By examining those shipping rates which are presented in tables 22 through 25,

The Transportation Models

the importance of the changes of the costs affecting optimum shipping flows is illustrated. As analyzed in previous chapters, ocean transportation cost is primarily determined by three factors: geographic distance between each pair of port of origin and port of destination, ship size and ship flag.

Generally speaking, as distance increases, freight rates are implicitly assumed to increase proportionately at a specific cost per unit. Therefore, total costs increase as ton-miles traveled increase. In the models, of all the seven selected ports of origin, port of Galveston is the nearest to any ports of destination. Therefore, shipping cost for shipment between port of Galveston and a certain destination port is the lowest among the costs of shipping between that destination port and other origin ports. And cost of shipping from port of Beaumont to a certain importing port is the highest among costs of shipping from other origin ports.

Size economies indicate that there exist cost advantages with the use of larger ships. Therefore, within the range of ship sizes in this study (25,000 DWT to 40,000 DWT), there is a difference in shipping costs between the larger and smaller ships with the larger ones having less cost per unit of ship ton. In the results of optimum shipping patterns determined in each model, 40,000 DWT vessels are chosen to be largely used due to their relatively lower costs.

As far as the flag of registry is concerned, of the three different flags used in the study, shipping costs for U.S.-flag ships are substantially higher than the costs of other ships. Liberia-flag ships have the lowest shipping costs. Therefore, in the optimum shipping patterns developed in each model, Liberia-flag ships are selected to be the optimal ships most commonly used for the shipments.

Actual Shipping Pattern

As shown in table 28, for the given surplus and deficit amounts in the seven ports of origin and thirteen destination areas, ports of New Orleans, Houston and Port Arthur were the principal ports of export that had shipments to almost all the importing markets. Ports of Mobile, Baton Rouge and Galveston were moderately used to ship rice to three different importing areas. Mobile shipped to regions of Central America and the Caribbean (438.4 tons), Northwestern and Central Europe (233 tons), and Western Africa (92.8 tons); while Baton Rouge exported to regions of Central

America and the Caribbean (4,726.9 tons), Northwestern and Central Europe (56,428.8 tons), and Southern Africa (3,617.5 tons); Galveston shipped to Eastern Southern America (44,707 tons), Western Asia (183,603.5 tons), and Eastern Africa (5,754.9 tons).

Table 28 Actual Shipping Pattern for U.S. Southern Region Rice Exports

To From	(1) CAC	(2) WSA	(3) ESA	(4) NCE	(5) NSE
			(in short ton)		
Mobile	438.4	0	0	233.0	0
Baton Rouge	4,726.9	0	0	56,428.8	0
New Orleans	46,099.6	202.0	22,400.5	74,770.1	443.3
Beaumont	0	0	0	0	0
Galveston	0	0	44,707.0	0	0
Houston	29,761.1	934.9	139,441.0	121,466.0	785.5
Port Arthur	100,980.8	7,458.7	401,236.3	70,366.5	20,814.0
Total	182,006.8	8,595.6	607,784.8	323,264.4	22,042.9

To From	(6) EA	(7) WA	(8) SSA	(9) AO	(10) WAF
			(in short ton)		
Mobile	0	0	0	0	92.8
Baton Rouge	0	0	0	0	0
New Orleans	202.0	86,950.2	2,781.2	1,544.9	5,013.9
Beaumont	0	0	0	0	146.3
Galveston	0	183,603.5	0	0	0
Houston	525.5	300,120.0	0	1,016.6	157,181.7
Port Arthur	0	67,287.2	676.5	0	91,564.7
Total	727.5	637,960.9	3,457.7	2,561.5	253,999.4

(continued on next page)

The Transportation Models

To From	(11) NAF	(12) EAF	(13) SAF	Total
		(in short ton)		
Mobile	0	0	0	764.2
Baton Rouge	0	0	3,617.5	64,773.2
New Orleans	203.8	724.5	19,287.4	260,623.5
Beaumont	0	0	0	146.3
Galveston	0	5,754.9	0	234,065.4
Houston	6.7	720.0	796.9	752,755.9
Port Arthur	101.0	78,179.0	62,619.6	901,284.3
Total	311.5	85,378.4	86,321.4	2,214,412.8

Source: U.S. Department of Commerce. Bureau of the Census. Unpublished Foreign Trade Statistics. Bureau of the Census, U.S. Department of Commerce, Washington, D.C. 1987.

Note: Column headings are as follows: (1) CAC: Central America & the Caribbean; (2) WSA: Western South America; (3) ESA: Eastern South America; (4) NCE: Northwestern & Central Europe; (5) NSE: Northeastern & Southern Europe; (6) EA: Eastern Asia; (7) WA: Western Asia; (8) SSA: Southeastern & Southern Asia; (9) AO: Australia & Oceania; (10) WAF: Western Africa; (11) NAF: Northern Africa; (12) EAF: Eastern Africa; and (13) SAF: Southern Africa.

Optimal Shipping Patterns.

It can be concluded from examining the total transportation cost of each model that there is no substantial difference among total costs in each model. The basic model has the lowest transportation cost which is the same for Models V and VI. Model II has the highest transportation cost which is over $1 million more than the cost in other models.

On an overall basis, a comparison of the results of all the six models and the actual shipping flow of rice in 1986 gave certain implications. In general, all the six models show similarities in routing patterns with least-cost which are different from the actual shipping pattern in 1986, with given assumptions and restrictions.

Model I

In this model, 40,000 DWT and Liberia-flag vessel is selected to be mostly used for carrying rice in the optimum shipping pattern, table 29. Almost all shipments are carried on 40,000 DWT and Liberia-flag vessels except for a few shipments that are made with 30,000 DWT, Japan-flag vessels at the cost of $6.1 and $5.3 per ton, respectively.

In the optimal shipping pattern, Port Arthur and port of Houston are most frequently used, from which rice shipments are carried to five market areas. The amounts shipped from those two ports are much larger than the shipments from other ports. Ports of Galveston and Baton Rouge are used less frequently. Galveston ships to Central America and the Caribbean, Australia and Oceania, Southern Africa and Eastern Asia region; Baton Rouge ships to Northwestern and Central Europe, and Southern Africa region. As previously shown in actual rice shipping patterns in 1986, these two ports shipped to three importing areas which are different from the shipments in the model.

From ports of Mobile, Beaumont and New Orleans, shipments are made to only one destination region. Port of New Orleans ships to Northwestern and Central Europe only. But in the actual shipping pattern in 1986, New Orleans was used as a principal port to ship rice to almost all deficit areas except Western South America and Eastern Asia. Port of Mobile exports rice to Northeastern and Southern Europe, with the amount being only 764.2 tons. In actual shipping pattern, port of Mobile shipped to three areas: Central America and the Caribbean, Northwestern and Central Europe, and Western Africa, with the same amount of 764.2 tons. With the optimal shipping pattern determined in this model, total transportation cost is $42,340,273.50 which is the lowest among all the models.

Model II

In this model, the transportation costs for the shipments are changed by increasing interest rates for optimal ships determined in Model I by 20 percent. With the adjusted transportation costs, the result of the model shows different optimum shipping patterns and a higher total cost of $43,666,216.50, table 30. The 40,000 DWT Liberia-flag ships are used for

The Transportation Models

almost all the shipments in this model. The 35,000 DWT and 30,000 DWT Liberia-flag ships are used several times. U.S.-flag and Japan-flag vessels are not used at all.

In the optimum shipping pattern determined by the model, Port Arthur exports the largest amount of rice to four regions. They are: Central America and the Caribbean, Western Asia, Southeastern and South Asia, and Eastern Africa. Port of Galveston ships to three deficit areas. They are: Eastern Asia, Northwestern and Central Europe, and Western Africa. Ports of Houston and Baton Rouge export to six and three regions, respectively, with the amount ships from Houston being the second largest, 752,756 tons. Port of Houston ships to Western South America, Eastern South America, Western Asia, Australia and Oceania, Western Africa and Southern Africa. Port of Baton Rouge ships to Northeastern and South Europe, Western Africa, and Northern Africa. Port of New Orleans exports to one area: Northwestern and Central Europe, for the moderate amount of 260,623.5 tons at the cost of $14.4 per ton. Port of Mobile ships to only one region, Western Africa. That port ships 764.2 tons at the cost of $14.30 per ton. In both Models I and II, Port Arthur is most frequently used for shipments of rice. Port of Houston is used more frequently in Model II (six regions) than in Model I (four regions).

Model III

In this model, the insurance rates for the optimal vessels determined in the Basic Model are increased by 20 percent so that the estimated shipping costs are adjusted for the optimum shipping pattern. The total transportation cost for the model is $42,571,701.50. As in Model I, the 40,000 DWT, Liberia-flag ships are utilized to export to almost all the deficit markets, table 31. The 30,000 DWT, Liberia-flag vessels are used to ship to Central America and the Caribbean region from Port Arthur for the amount of 182,006.8 tons at the cost of $6.1 per ton.

In the optimal shipping pattern of this model, Port Arthur exports to seven deficit areas for the largest amount of 901,284.3 tons. This result is the same as in Model I. These areas are: Central America and the Caribbean, Western South America, Northwestern and Central Europe, Eastern Asia, Western Asia, Eastern Africa, and Western Africa.

Ports of Houston and Baton Rouge each ships to three and four regions, while in Model I, port of Houston ships to five deficit regions. In this model, port of Houston ships to Eastern South America, Western Asia, Southeastern and South Asia, Australia and Oceania, and Western Africa; Baton Rouge ships to Eastern South America, Northeastern and Southern Europe, and Northern Africa. Port of New Orleans ships to Eastern Southern America for the amount of 260,623.5 tons, and port of Mobile ships to Southern Africa only for the amount of $764.2 tons at the expense of $19.3 per ton. This is similar to that in Model I.

Model IV

In this model, port expenses in the ports of New Orleans and Houston are increased by 20 percent in order to change the transportation cost. The total transportation cost for the optimal shipping pattern in this model is $42,340,460.50. Most shipments are carried on 40,000 DWT, Liberia-flag ships, which is the same as in Model I, table 32. The amount of 182,006.8 tons is shipped on 30,000 DWT, Liberia-flag ships from Port Arthur and Galveston to Central America and the Caribbean area at the cost of $6.1 and $5.3 per ton. The 30,000 DWT, Japan-flag ships are used to carry 311.5 tons of rice from Port Arthur to Northern Africa at the cost of $13.8 per ton.

In the optimum shipping pattern, exports to five rice deficit markets are shipped from Port Arthur for the amount of 901,284.3 tons which is the largest amount among all the ports of origin as in Model I. Ports of Houston and New Orleans each exports to two areas, while in Model I, Houston ships to five deficit areas. Houston ships to Western Asia and Western Africa in this model; New Orleans ships to Northwestern and Central Europe, and Southern Africa. Port of Baton Rouge ships to one market area. This is similar to the result of Model I.

In this model, port of Baton Rouge ships to Northwestern and Central Europe (64,773.2 tons); port of Galveston ships to seven areas. Port of Mobile exports to one region: Northeastern and Southern Europe for the amount of 764.2 tons at the cost of $14.1 per ton. This is the same as in Model I.

Table 29 Model I: Optimal Shipping Pattern for U.S. Rice Exports

To From	(1) CAC	(2) WSA	(3) ESA	(4) NCE	(5) NSE
Mobile					
Short ton	0	0	0	0	764.2
Cost: $/ton					14.1
Flag/Size: DWT					Liberia/40
Baton Rouge					
Short ton	0	0	0	62,640.9	0
Cost: $/ton				14.3	
Flag/Size				Liberia/40	
New Orleans					
Short ton	0	0	0	260,623.5	0
Cost: $/ton				14.0	
Flag/Size				Liberia/40	
Beaumont					
Short ton	0	0	0	0	0
Cost: $/ton					
Flag/Size					
Galveston					
Short ton	145,398.0	0	0	0	0
Cost: $/ton	5.3				
Flag/Size	Liberia/30				
Houston					
Short ton	0	8,595.6	0	0	21,278.7
Cost: $/ton		11.1			15.4
Flag/Size		Liberia/40			Liberia/40
Port Arthur					
Short ton	36,608.8	0	607,784.8	0	0
Cost: $/ton	6.1		15.1		
Flag/Size	Liberia/30		Liberia/40		
Total	182,006.8	8,595.6	607,784.8	323,264.4	22,042.9

(continued on next page)

To From	(6) EA	(7) WA	(8) SSA	(9) AO	(10) WAF
Mobile					
Short ton	0	0	0	0	0
Cost: $/ton					
Flag/Size					
Baton Rouge					
Short ton	0	0	0	0	0
Cost: $/ton					
Flag/Size					
New Orleans					
Short ton	0	0	0	0	0
Cost: $/ton					
Flag/Size					
Beaumont					
Short ton	0	0	146.3	0	0
Cost: $/ton			29.1		
Flag/Size			Liberia/35		
Galveston					
Short ton	727.5	0	0	2,561.5	0
Cost: $/ton	18.9			21.2	
Flag/Size	Liberia/40			Liberia/40	
Houston					
Short ton	0	637,960.9	0	0	84,609.2
Cost: $/ton		30.6			15.1
Flag/Size		Liberia/40			Liberia/40
Port Arthur					
Short ton	0	0	3,311.4	0	169,390.2
Cost: $/ton			27.9		14.6
Flag/Size			Liberia/40		Liberia/40
Total	727.5	637,960.9	3,457.7	2,561.5	253,999.4

(continued on next page)

The Transportation Models

To From	(11) NAF	(12) EAF	(13) SAF	Total
Mobile				
Short ton	0	0	0	764.2
Cost: $/ton				
Flag/Size				
Baton Rouge				
Short ton	0	0	2,132.3	64,773.2
Cost: $/ton			19.6	
Flag/Size			Liberia/40	
New Orleans				
Short ton	0	0	0	260,623.5
Cost: $/ton				
Flag/Size				
Beaumont				
Short ton	0	0	0	146.3
Cost: $/ton				
Flag/Size				
Galveston				
Short ton	0	85,378.4	0	234,065.4
Cost: $/ton		23.8		
Flag/Size		Liberia/40		
Houston				
Short ton	311.5	0	0	752,755.9
Cost: $/ton	13.7			
Flag/Size	Japan/30			
Port Arthur				
Short ton	0	0	84,189.1	901,284.3
Cost: $/ton			19.9	
Flag/Size			Liberia/40	
Total	311.5	85,378.4	86,321.4	2,214,412.8

Note: Column headings are as follows: (1) CAC: Central America & the Caribbean; (2) WSA: Western South America; (3) ESA: Eastern South America; (4) NCE: Northwestern & Central Europe; (5) NSE: Northeastern & Southern Europe; (6) EA: Eastern Asia; (7) WA: Western Asia; (8) SSA: Southeastern & Southern Asia; (9) AO: Australia & Oceania; (10) WAF: Western Africa; (11) NAF: Northern Africa; (12) EAF: Eastern Africa; and (13) SAF: Southern Africa.

Table 30 Model II: Optimal Shipping Pattern for U.S. Rice Exports

To From	(1) CAC	(2) WSA	(3) ESA	(4) NCE	(5) NSE
Mobile					
Short ton	0	0	0	0	0
Cost: $/ton					
Flag/Size: 000DWT					
Baton Rouge					
Short ton	0	0	0	0	22,042.9
Cost: $/ton					15.0
Flag/Size					Liberia/40
New Orleans					
Short ton	0	0	0	260,623.5	0
Cost: $/ton				14.4	
Flag/Size				Liberia/40	
Beaumont					
Short ton	0	0	0	0	0
Cost: $/ton					
Flag/Size					
Galveston					
Short ton	0	0	0	62,640.9	0
Cost: $/ton				14.2	
Flag/Size				Liberia/40	
Houston					
Short ton	0	8,595.6	607,784.8	0	0
Cost: $/ton		11.5	16.2		
Flag/Size		Liberia/35	Liberia/40		
Port Arthur					
Short ton	182,006.8	0	0	0	0
Cost: $/ton	6.1				
Flag/Size	Liberia/30				
Total	182,006.8	8,595.6	607,784.8	323,264.4	22,042.9

(continued on next page)

The Transportation Models 119

To From	(6) EA	(7) WA	(8) SSA	(9) AO	(10) WAF
Mobile					
Short ton	0	0	0	0	764.2
Cost: $/ton					14.3
Flag/Size					Liberia/40
Baton Rouge					
Short ton	0	0	0	0	42,418.8
Cost: $/ton					14.8
Flag/Size					Liberia/40
New Orleans					
Short ton	0	0	0	0	0
Cost: $/ton					
Flag/Size					
Beaumont					
Short ton	0	0	146.3	0	0
Cost: $/ton			29.1		
Flag/Size			Liberia/35		
Galveston					
Short ton	727.5	0	0	0	170,697.0
Cost: $/ton	19.5				14.2
Flag/Size	Liberia/40				Liberia/40
Houston					
Short ton	0	7,373.2	0	2,561.5	40,119.4
Cost: $/ton		31.6		23.3	15.6
Flag/Size		Liberia/40		Liberia/40	Liberia/40
Port Arthur					
Short ton	0	630,587.7	3,311.4	0	0
Cost: $/ton		31.0	28.8		
Flag/Size		Liberia/40	Liberia/40		
Total	727.5	637,960.9	3,457.7	2,561.5	253,999.4

(*continued on next page*)

To From	(11) NAF	(12) EAF	(13) SAF	Total
Mobile				
Short ton	0	0	0	764.2
Cost: $/ton				
Flag/Size				
Baton Rouge				
Short ton	311.5	0	0	64,773.2
Cost: $/ton	15.1			
Flag/Size	Liberia/40			
New Orleans				
Short ton	0	0	0	260,623.5
Cost: $/ton				
Flag/Size				
Beaumont				
Short ton	0	0	0	146.3
Cost: $/ton				
Flag/Size				
Galveston				
Short ton	0	0	0	234,065.4
Cost: $/ton				
Flag/Size				
Houston				
Short ton	0	0	86,321.4	752,755.9
Cost: $/ton			21.1	
Flag/Size			Liberia/40	
Port Arthur				
Short ton	0	85,378.4	0	901,284.3
Cost: $/ton		25.4		
Flag/Size		Liberia/40		
Total	311.5	85,378.4	86,321.4	2,214,412.8

Note: Column headings are as follows: (1) CAC: Central America & the Caribbean; (2) WSA: Western South America; (3) ESA: Eastern South America; (4) NCE: Northwestern & Central Europe; (5) NSE: Northeastern & Southern Europe; (6) EA: Eastern Asia; (7) WA: Western Asia; (8) SSA: Southeastern & Southern Asia; (9) AO: Australia & Oceania; (10) WAF: Western Africa; (11) NAF: Northern Africa; (12) EAF: Eastern Africa; and (13) SAF: Southern Africa.

The Transportation Models

Table 31 Model III: Optimal Shipping Pattern for U.S. Rice Exports

To From	(1) CAC	(2) WSA	(3) ESA	(4) NCE	(5) NSE
Mobile					
Short ton	0	0	0	0	0
Cost: $/ton					
Flag/Size: 000DWT					
Baton Rouge					
Short ton	0	0	42,418.8	0	22,042.9
Cost: $/ton			15.0		14.6
Flag/Size			Liberia/40		Liberia/40
New Orleans					
Short ton	0	0	260,623.5	0	0
Cost: $/ton			14.6		
Flag/Size			Liberia/40		
Beaumont					
Short ton	0	0	0	0	0
Cost: $/ton					
Flag/Size					
Galveston					
Short ton	0	0	148,508.2	0	0
Cost: $/ton			14.4		
Flag/Size			Liberia/40		
Houston					
Short ton	0	0	156,234.3	0	0
Cost: $/ton			15.8		
Flag/Size			Liberia/40		
Port Arthur					
Short ton	182,006.8	8,595.6	0	323,264.4	0
Cost: $/ton	16.1	10.6		14.6	
Flag/Size	Liberia/30	Liberia/40		Liberia/40	
Total	182,006.8	8,595.6	607,784.8	323,264.4	22,042.9

(continued on next page)

To From	(6) EA	(7) WA	(8) SSA	(9) AO	(10) WAF
Mobile					
Short ton	0	0	0	0	0
Cost: $/ton					
Flag/Size					
Baton Rouge					
Short ton	0	0	0	0	0
Cost: $/ton					
Flag/Size					
New Orleans					
Short ton	0	0	0	0	0
Cost: $/ton					
Flag/Size					
Beaumont					
Short ton	0	0	146.3	0	0
Cost: $/ton			29.1		
Flag/Size			Liberia/35		
Galveston					
Short ton	0	0	0	0	0
Cost: $/ton					
Flag/Size					
Houston					
Short ton	0	590,648.7	3,311.4	2,561.5	0
Cost: $/ton		30.8	28.7	22.7	
Flag/Size		Liberia/40	Liberia/40	Liberia/40	
Port Arthur					
Short ton	727.5	47,312.2	0	0	253,999.4
Cost: $/ton	19.8	30.2			14.6
Flag/Size	Liberia/40	Liberia/40			Liberia/40
Total	727.5	637,960.9	3,457.7	2,561.5	253,999.4

(continued on next page)

To From	(11) NAF	(12) EAF	(13) SAF	Total
Mobile				
Short ton	0	0	764.2	764.2
Cost: $/ton			19.3	
Flag/Size			Liberia/40	
Baton Rouge				
Short ton	311.5	0	0	64,773.2
Cost: $/ton	14.7			
Flag/Size	Liberia/40			
New Orleans				
Short ton	0	0	0	260,623.5
Cost: $/ton				
Flag/Size				
Beaumont				
Short ton	0	0	0	146.3
Cost: $/ton				
Flag/Size				
Galveston				
Short ton	0	0	85,557.2	234,065.4
Cost: $/ton			19.2	
Flag/Size			Liberia/40	
Houston				
Short ton	0	0	0	752,755.9
Cost: $/ton				
Flag/Size				
Port Arthur				
Short ton	0	85,378.4	0	901,284.3
Cost: $/ton		24.8		
Flag/Size		Liberia/40		
Total	311.5	85,378.4	86,321.4	2,214,412.8

Note: Column headings are as follows: (1) CAC: Central America & the Caribbean; (2) WSA: Western South America; (3) ESA: Eastern South America; (4) NCE: Northwestern & Central Europe; (5) NSE: Northeastern & Southern Europe; (6) EA: Eastern Asia; (7) WA: Western Asia; (8) SSA: Southeastern & Southern Asia; (9) AO: Australia & Oceania; (10) WAF: Western Africa; (11) NAF: Northern Africa; (12) EAF: Eastern Africa; and (13) SAF: Southern Africa.

Table 32 Model IV: Optimal Shipping Pattern for U.S. Rice Exports

To From	(1) CAC	(2) WSA	(3) ESA	(4) NCE	(5) NSE
Mobile					
Short ton	0	0	0	0	764.2
Cost: $/ton					14.1
Flag/Size: 000DWT					Liberia/40
Baton Rouge					
Short ton	0	0	0	64,773.3	0
Cost: $/ton				14.3	
Flag/Size				Liberia/40	
New Orleans					
Short ton	0	0	0	174,302.1	0
Cost: $/ton				14.0	
Flag/Size				Liberia/40	
Beaumont					
Short ton	0	0	0	0	0
Cost: $/ton					
Flag/Size					
Galveston					
Short ton	31,334.5	8,595.6	0	84,189.1	21,278.7
Cost: $/ton	5.3	9.8		13.8	14.1
Flag/Size	Liberia/30	Liberia/40		Liberia/40	Liberia/40
Houston					
Short ton	0	0	0	0	0
Cost: $/ton					
Flag/Size					
Port Arthur					
Short ton	150,672.3	0	607,784.8	0	0
Cost: $/ton	6.1		15.1		
Flag/Size	Liberia/30		Liberia/40		
Total	182,006.8	8,595.6	607,784.8	323,264.4	22,042.9

(continued on next page)

To From	(6) EA	(7) WA	(8) SSA	(9) AO	(10) WAF
Mobile					
Short ton	0	0	0	0	0
Cost: $/ton					
Flag/Size					
Baton Rouge					
Short ton	0	0	0	0	0
Cost: $/ton					
Flag/Size					
New Orleans					
Short ton	0	0	0	0	0
Cost: $/ton					
Flag/Size					
Beaumont					
Short ton	0	0	146.3	0	0
Cost: $/ton			29.1		
Flag/Size			Liberia/35		
Galveston					
Short ton	727.5	0	0	2,561.5	0
Cost: $/ton	18.9			21.2	
Flag/Size	Liberia/40			Liberia/40	
Houston					
Short ton	0	637,960.9	0	0	114,795.0
Cost: $/ton		30.6			15.1
Flag/Size		Liberia/40			Liberia/40
Port Arthur					
Short ton	0	0	3,311.4	0	139,204.4
Cost: $/ton			27.9		14.6
Flag/Size			Liberia/40		Liberia/40
Total	727.5	637,960.9	3,457.7	2,561.5	253,999.4

(*continued on next page*)

To From	(11) NAF	(12) EAF	(13) SAF	Total
Mobile				
Short ton	0	0	0	764.2
Cost: $/ton				
Flag/Size				
Baton Rouge				
Short ton	0	0	0	64,773.2
Cost: $/ton				
Flag/Size				
New Orleans				
Short ton	0	0	86,321.4	260,623.5
Cost: $/ton			19.3	
Flag/Size			Liberia/40	
Beaumont				
Short ton	0	0	0	146.3
Cost: $/ton				
Flag/Size				
Galveston				
Short ton	0	85,378.4	0	234,065.4
Cost: $/ton		23.8		
Flag/Size		Liberia/40		
Houston				
Short ton	0	0	0	752,755.9
Cost: $/ton				
Flag/Size				
Port Arthur				
Short ton	311.5	0	0	901,284.3
Cost: $/ton	13.8			
Flag/Size	Japan/30			
Total	311.5	85,378.4	86,321.4	2,214,412.8

Note: Column headings are as follows: (1) CAC: Central America & the Caribbean; (2) WSA: Western South America; (3) ESA: Eastern South America; (4) NCE: Northwestern & Central Europe; (5) NSE: Northeastern & Southern Europe; (6) EA: Eastern Asia; (7) WA: Western Asia; (8) SSA: Southeastern & Southern Asia; (9) AO: Australia & Oceania; (10) WAF: Western Africa; (11) NAF: Northern Africa; (12) EAF: Eastern Africa; and (13) SAF: Southern Africa.

Model V

In this model, transportation costs are adjusted by reducing the operating costs of U.S.-flag vessels by 50 percent. The optimal shipping pattern shows the same total cost as the Basic Model: $42,340,273.50. Most shipments are made on the 40,000 DWT, Liberia-flag ships that had lower shipping costs, table 33. This is the same as in Model I. The 30,000 DWT, Liberia-flag vessels are used to ship 182,006.8 tons of rice from Port Arthur to Central America and the Caribbean regions. In this model, 311.5 tons of rice are shipped on the 30,000 DWT, Japan-flag vessels from port of Houston to Northern Africa at the cost of $13.7 per ton.

In the optimum shipping pattern of the model, like in Model I, port of Houston exports to five deficit regions for the amount of 752,755.9 tons. These regions are: Northeastern and Southern Europe, Western Asia, Australia and Oceania, Western Africa and Northern Africa. Port Arthur ships to five areas: Central America and the Caribbean, Eastern South America, Western Asia, Southeastern and Southern Asia and Southern Africa. This is also the same as in Model I. Port of Galveston exports to four market regions for moderate amounts. And ports of Mobile and Baton Rouge export to only one deficit area, Europe region, which is similar to the result in Model I.

Model VI

In this model, the construction costs of building U.S.-flag ships are reduced by 50 percent in order to adjust the transportation costs. The total shipping cost for the optimal shipping flows in this model is again the lowest: $42,340,273.50, which is the same as Model I. The 40,000 DWT, Liberia-flag ships are again used to carry most shipments, table 34. The 30,000 DWT, Liberia-flag ships are used to ship the amount of 182,006.8 tons from port of Galveston to the area of Central America and the Caribbean at the cost of $5.30 per ton. An amount of 311.5 tons is carried on the 30,000 DWT, Japan-flag vessels from the port of Houston to Northern Africa at the shipping cost of $13.7 per ton.

In the optimum shipping pattern of this model, Port Arthur exports to five rice deficit world markets for the largest amount: Eastern South America, Eastern Asia, Western Asia, Southeastern and Southern Asia and Eastern Africa. In the shipping pattern of Model I, Port Arthur ships to five deficit regions. Port of Houston ships to seven areas for the second largest amount in this model: Northeastern and Southern Europe, Western Asia, Australia and Oceania, Western Africa, Northern Africa, Western Southern America and Southern Africa, while in Model I, this port ships to five areas. Port of Galveston ships to two regions for the third largest amount: Central America and the Caribbean, and Western Asia. Ports of Mobile and New Orleans ship to only one area, Europe. This is similar to result in Model I.

SUMMARY

The objective of this chapter is to develop representative models to test and analyze the effects of transportation costs estimated in Chapter III on the flow of rice shipments in international trade. Since cost minimization is selected as the primary goal of this study, six transportation models, within linear programming framework, are employed to determine optimal flows of rice shipments, with different shipping costs, among U.S. exporting ports and world importing regions which would minimize the total costs of transportation.

The surplus amount for U.S. southern exporting regions and the deficit amount for the world importing regions are based on 1986 U.S. rice shipment data. The Basic Model (Model I) is developed to determine the optimal shipping pattern for U.S. 1986 rice shipment with minimized total transportation cost. The other five models' adjusted transportation costs are used to estimate optimum shipping patterns for different shipping costs. A comparison of the basic model and the actual shipping pattern in 1986 is conducted. There are several assumptions for the linear programming technique, based on which the six models are established.

The Basic Model develops an optimal shipping pattern with least transportation cost which is the lowest among all the models. The other five models that are subjected to adjusted transportation costs bear similarities in the optimal shipping patterns with the first model. They are different from the actual shipping pattern in 1986. Of all the U.S. southern ports of origin, ports of Houston and Port Arthur are used most frequently for exporting rice to the world markets in all the models.

Table 33 Model V: Optimal Shipping Pattern for U.S. Rice Exports

To From	(1) CAC	(2) WSA	(3) ESA	(4) NCE	(5) NSE
Mobile					
Short ton	0	0	0	0	764.2
Cost: $/ton					14.1
Flag/Size: 000DWT					Liberia/40
Baton Rouge					
Short ton	0	0	0	64,773.3	0
Cost: $/ton				14.3	
Flag/Size				Liberia/40	
New Orleans					
Short ton	0	0	0	258,491.2	0
Cost: $/ton				14.0	
Flag/Size				Liberia/40	
Beaumont					
Short ton	0	0	0	0	0
Cost: $/ton					
Flag/Size					
Galveston					
Short ton	0	8,595.6	0	0	0
Cost: $/ton		9.8			
Flag/Size		Liberia/40			
Houston					
Short ton	0	0	0	0	21,278.7
Cost: $/ton					15.4
Flag/Size					Liberia/40
Port Arthur					
Short ton	182,006.8	0	607,784.8	0	0
Cost: $/ton	6.1		15.1		
Flag/Size	Liberia/30		Liberia/40		
Total	182,006.8	8,595.6	607,784.8	323,264.4	22,042.9

(continued on next page)

To From	(6) EA	(7) WA	(8) SSA	(9) AO	(10) WAF
Mobile					
Short ton	0	0	0	0	0
Cost: $/ton					
Flag/Size					
Baton Rouge					
Short ton	0	0	0	0	0
Cost: $/ton					
Flag/Size					
New Orleans					
Short ton	0	0	0	0	0
Cost: $/ton					
Flag/Size					
Beaumont					
Short ton	0	0	146.3	0	0
Cost: $/ton			29.1		
Flag/Size			Liberia/35		
Galveston					
Short ton	727.5	140,553.2	0	0	0
Cost: $/ton	18.9	29.3			
Flag/Size	Liberia/40	Liberia/40			
Houston					
Short ton	0	474,604.8	0	2,561.5	253,999.4
Cost: $/ton		30.6		22.5	15.1
Flag/Size		Liberia/40		Liberia/40	Liberia/40
Port Arthur					
Short ton	0	22,802.9	3,311.4	0	0
Cost: $/ton		30.1	27.9		
Flag/Size		Liberia/40	Liberia/40		
Total	727.5	637,960.9	3,457.7	2,561.5	253,999.4

(*continued on next page*)

To From	(11) NAF	(12) EAF	(13) SAF	Total
Mobile				
Short ton	0	0	0	764.2
Cost: $/ton				
Flag/Size				
Baton Rouge				
Short ton	0	0	0	64,773.2
Cost: $/ton				
Flag/Size				
New Orleans				
Short ton	0	0	2,132.3	260,623.5
Cost: $/ton			19.3	
Flag/Size			Liberia/40	
Beaumont				
Short ton	0	0	0	146.3
Cost: $/ton				
Flag/Size				
Galveston				
Short ton	0	0	84,189.1	234,065.4
Cost: $/ton			19.1	
Flag/Size			Liberia/40	
Houston				
Short ton	311.5	0	0	752,755.9
Cost: $/ton	13.7			
Flag/Size	Japan/30			
Port Arthur				
Short ton	0	85,378.4	0	901,284.3
Cost: $/ton		24.6		
Flag/Size		Liberia/40		
Total	311.5	85,378.4	86,321.4	2,214,412.8

Note: Column headings are as follows: (1) CAC: Central America & the Caribbean; (2) WSA: Western South America; (3) ESA: Eastern South America; (4) NCE: Northwestern & Central Europe; (5) NSE: Northeastern & Southern Europe; (6) EA: Eastern Asia; (7) WA: Western Asia; (8) SSA: Southeastern & Southern Asia; (9) AO: Australia & Oceania; (10) WAF: Western Africa; (11) NAF: Northern Africa; (12) EAF: Eastern Africa; and (13) SAF: Southern Africa.

Table 34 Model VI: Optimal Shipping Pattern for U.S. Rice Exports

To From	(1) CAC	(2) WSA	(3) ESA	(4) NCE	(5) NSE
Mobile					
Short ton	0	0	0	0	764.2
Cost: $/ton					14.1
Flag/Size: 000DWT					Liberia/40
Baton Rouge					
Short ton	0	0	0	62,640.9	0
Cost: $/ton				14.3	
Flag/Size				Liberia/40	
New Orleans					
Short ton	0	0	0	260,623.5	0
Cost: $/ton				14.0	
Flag/Size				Liberia/40	
Beaumont					
Short ton	0	0	0	0	0
Cost: $/ton					
Flag/Size					
Galveston					
Short ton	182,006.8	0	0	0	0
Cost: $/ton	5.3				
Flag/Size	Liberia/40				
Houston					
Short ton	0	8,595.6	0	0	21,278.7
Cost: $/ton		11.1			15.4
Flag/Size		Liberia/40			Liberia/40
Port Arthur					
Short ton	0	0	607,784.8	0	0
Cost: $/ton			15.1		
Flag/Size			Liberia/40		
Total	182,006.8	8,595.6	607,784.8	323,264.4	22,042.9

(continued on next page)

The Transportation Models

To From	(6) EA	(7) WA	(8) SSA	(9) AO	(10) WAF
Mobile					
Short ton	0	0	0	0	0
Cost: $/ton					
Flag/Size					
Baton Rouge					
Short ton	0	2,132.3	0	0	0
Cost: $/ton		29.8			
Flag/Size		Liberia/40			
New Orleans					
Short ton	0	0	0	0	0
Cost: $/ton					
Flag/Size					
Beaumont					
Short ton	0	0	146.3	0	0
Cost: $/ton			29.1		
Flag/Size			Liberia/35		
Galveston					
Short ton	0	52,058.6	0	0	0
Cost: $/ton		29.3			
Flag/Size		Liberia/40			
Houston					
Short ton	0	379,687.8	0	2,561.5	253,999.4
Cost: $/ton		30.6		22.5	15.1
Flag/Size		Liberia/40		Liberia/40	Liberia/40
Port Arthur					
Short ton	727.5	204,082.2	3,311.4	0	0
Cost: $/ton	19.7	30.1	27.9		
Flag/Size	Liberia/40	Liberia/40	Liberia/40		
Total	727.5	637,960.9	3,457.7	2,561.5	253,999.4

(*continued on next page*)

To From	(11) NAF	(12) EAF	(13) SAF	Total
Mobile				
Short ton	0	0	0	764.2
Cost: $/ton				
Flag/Size				
Baton Rouge				
Short ton	0	0	0	64,773.2
Cost: $/ton				
Flag/Size				
New Orleans				
Short ton	0	0	0	260,623.5
Cost: $/ton				
Flag/Size				
Beaumont				
Short ton	0	0	0	146.3
Cost: $/ton				
Flag/Size				
Galveston				
Short ton	0	0	0	234,065.4
Cost: $/ton				
Flag/Size				
Houston				
Short ton	311.5	0	86,321.4	752,755.9
Cost: $/ton	13.7		20.4	
Flag/Size	Japan/30		Liberia/40	
Port Arthur				
Short ton	0	85,378.4	0	901,284.3
Cost: $/ton		24.6		
Flag/Size		Liberia/40		
Total	311.5	85,378.4	86,321.4	2,214,412.8

Note: Column headings are as follows: (1) CAC: Central America & the Caribbean; (2) WSA: Western South America; (3) ESA: Eastern South America; (4) NCE: Northwestern & Central Europe; (5) NSE: Northeastern & Southern Europe; (6) EA: Eastern Asia; (7) WA: Western Asia; (8) SSA: Southeastern & Southern Asia; (9) AO: Australia & Oceania; (10) WAF: Western Africa; (11) NAF: Northern Africa; (12) EAF: Eastern Africa; and (13) SAF: Southern Africa.

NOTES

1. Frank Hitchcock. *The Distribution of a Product From Several Sources to Numerous Localities*. Journal of Mathematics and Physics, 20. 1941.
2. G.B. Dantzig. *Activity Analysis of Production and Allocation*. Cowles Commission Monograph 13. 1951.
3. A Charnes and W.W. Cooper. *The Stepping Stone Method of Explaining Linear Programming Calculations in Transportation Problems*. Management Science 1:49-69. 1954.
4. M. LeRoy Davis. *Cost of Shipping United States Grain Exports to Principal World Markets*. Unpublished Master Thesis, Department of Agricultural Economics, Iowa State University. 1968.
5. U.S. Department of Commerce, Bureau of the Census. *Unpublished Foreign Trade Statistics*, SA705/705IT, 1986. Bureau of the Census, U.S. Department of Commerce, Washington, D.C., 1986.
6. Musa Pinar. *Analysis of Ocean Transportation Costs and Tariff Barriers in International Cotton Trade*. Unpublished Ph.D. dissertation, Mississippi State University. December 1983.
7. Earl O. Heady and Wilfred Candler. Linear Programming Methods. Iowa State University Press, Ames, Iowa. 1958.

V

Summary And Conclusions

SUMMARY

Over the years, numerous studies have been conducted to analyze international trade, including rice trade. However, few of these studies incorporated the effects of transportation costs in the analysis. There is not much research to date that attempts to examine the influence of transportation costs on the flow of international rice trade.

The analysis in this book indicates that international trade depends not only on the demand and supply conditions, but also on the "trade resistance" factors such as transportation costs and other restrictions among the trading regions. It is believed that these factors also play an important role in determining the level and direction of trade. The importance of transportation costs to the world economy has been increasingly realized since they influence substantially the directions and magnitudes of trade flows and gains from trade.

The United States is one of the major rice producing and exporting countries in the world. Rice is a major commodity for the United States in international trade. The U.S. rice industry has been facing the question of how to keep an appropriate share of the world market. The southern region of the United States is a major rice producing and exporting area, and includes the states of Arkansas, Louisiana, Mississippi and Texas.

The cost of transportation service rates on shipping rice through various kinds of vessels from U.S. southern region to international markets is one of the major sources of trade restrictions for U.S. rice exports. Any change in rice transportation costs will have a substantial impact on rice production, rice exports and other factors involved in the movement of rice from that region to importing countries, and thus will severely affect the competition of the U.S. rice industry in the international rice market. Lower transportation costs will enable U.S. rice producers and exporters to maintain and improve their competitive position in the world market.

Since changes in ocean transportation costs are very important in determining the routes over which rice should be shipped, and the export routes that are determined by shipping costs affect the routing of rice from the U.S. southern region to world market, the ocean transportation costs have an indirect effect on U.S. rice economy.

Therefore, the general objective of this study is to determine the cost per ton of shipping rice for four sizes of bulk rice vessels from U.S. southern ports of origin to specific points of foreign importing ports. These cost data are then used in constructing transportation models to determine least-cost optimal shipping patterns for U.S. rice exports.

First, some background information concerning rice shipments are developed. There are generally three types of ships used for carrying rice in exports: cargo liners, tanker vessels and tramp steamers. Liners travel fixed routes in a regular scheduled service, according to a predetermined time table at set rates. Tanker vessels usually carry large tonnages of single commodities by operating one or a fleet of ships especially designed for one cargo. They are infrequently used in shipping rice. Tramp steamers travel no predetermined route. They are by far the most frequently used vessels for carrying rice exports from the U.S. to international markets.

U.S. fleet vessels carry either U.S.-flag or foreign-flag. There has been a large difference in unit transportation costs between the two, with the U.S.-flag vessels operating at much higher costs than the foreign-flag vessels. In this study, three different flags have been analyzed: U.S.-flag, Japan-flag and Liberia-flag, with Liberia-flag vessels having the lowest shipping cost. U.S.-flag ships, even with higher capital cost, higher wage cost and higher fuel cost, are able to compete in some trade with their foreign-counterparts due to government subsidization.

Second, the types of costs involved in the transportation of shipping rice from U.S. ports of origin to foreign importing ports are analyzed. Ocean transportation cost is simply the price (cost) paid for transferring goods from point of export to point of import. Costs encountered in owning and operating a vessel can be generally classed under three categories: (1) vessel ownership expenses, (2) at-sea expenses and (3) in-port expenses. Shipping distance, ship size and trade volume are the most important factors affecting ocean transportation costs. Freight rates are assumed to increase

Summary and Conclusions

proportionately at a specific cost per unit as ton-miles traveled increase. Size economies imply that larger ships have lower cost than the smaller ones. Four types of sizes are used in the analysis: 25,000 DWT, 30,000 DWT, 35,000 DWT and 40,000 DWT. Along a particular route, areas with large volumes of trade have relatively lower terminal cost per ton as compared to areas with smaller volumes of trade.

Thirdly, the cost per ton of shipping rice from selected points of U.S. southern ports to world markets is estimated. As stated before, ocean transportation costs are an integral part of the data required to develop transportation models in order to find optimal shipping routes for U.S. rice exports with least costs. Transportation cost per ton for basic model can be approximated by the following equations for each of the twelve ships studied:

(1) 25,000 DWT, U.S.-flag Ship
T.C./ton = 1.959 (days in port) + 2.323 (days at sea) + 0.061

(2) 25,000 DWT, Japan-flag Ship
T.C./ton = 0.943 (days in port) + 1.307 (days at sea) + 0.061

(3) 25,000 DWT, Liberia-flag Ship
T.C./ton = 0.714 (days in port) + 1.078 (days at sea) + 0.061

(4) 30,000 DWT, U.S.-flag Ship
T.C./ton = 1.838 (days in port) + 2.14 (days at sea) + 0.061

(5) 30,000 DWT, Japan-flag Ship
T.C./ton = 0.865 (days in port) + 1.167 (days at sea) + 0.061

(6) 30,000 DWT, Liberia-flag Ship
T.C./ton = 0.673 (days in port) + 0.975 (days at sea) + 0.061

(7) 35,000 DWT, U.S.-flag Ship
T.C./ton = 1.753 (days in port) + 2.009 (days at sea) + 0.055

(8) 35,000 DWT, Japan-flag Ship
T.C./ton = 0.810 (days in port) + 1.066 (days at sea) + 0.055

(9) 35,000 DWT, Liberia-flag Ship

T.C./ton = 0.646 (days in port) + 0.902 (days at sea) + 0.055

(10) 40,000 DWT, U.S.-flag Ship
T.C./ton = 1.688 (days in port) + 1.910 (days at sea) + 0.056

(11) 40,000 DWT, Japan-flag Ship
T.C./ton = 0.768 (days in port) + 0.991 (days at sea) + 0.056

(12) 40,000 DWT, Liberia-flag Ship
T.C./ton = 0.625 (days in port) + 0.847 (days at sea) + 0.056

Unit transportation rates per ton for each of the twelve ships are calculated from the above relationships for Model I. Days in port are obtained from contacting the selected U.S. port authorities. Days at sea are estimated by dividing the distance between ports by the number of miles traveled per day by each ship. Transportation rates for the other five models are adjusted through the similar procedure. Transportation models can then be used to describe and predict rice flows among origins and destinations.

Fourth, transportation models are constructed to derive least cost shipping patterns for U.S. southern region rice exports under different conditions. In this study, cost minimization is selected as the primary goal, and therefore six transportation models, within linear programming framework, are employed to determine the optimal flows of rice shipments among exporting regions and importing countries which minimize the total costs of transportation.

Data of rice trade are collected for the period 1981-1987, and the 1986 calendar year is chosen as the data base for the study. That is, the amount available for export from each port of origin and the distribution of exports among deficit areas of the world are estimated from 1986 United States rice exports. The U.S. southern region rice exporting ports are grouped into seven surplus regions with each having a base point through which all the shipments would be made. Likewise, countries importing rice from U.S. are grouped into 13 deficit regions and a base point is chosen within each region through which all the shipments would occur, For comparison purposes, the actual shipping routes for rice exports from the seven selected U.S. southern region to the 13 importing areas in 1986 are calculated and presented in table 28.

Based on the basic assumptions of linear programming, six transportation models are developed by utilizing FMPS technique. For

Summary and Conclusions

analytical purposes, one basic model (Model I) is set up to determine an optimal shipping pattern of rice exports from the U.S. southern region to world markets by minimizing the total transportation costs with the given data and estimated unit shipping costs.

Five additional models are constructed and examined with different transportation rates. The second model assumed that for the optimal ships determined in Model I, interest rates in estimating shipping cost are increased by 20 percent. In Model III, for the optimal ships determined in the basic model, the insurance rates are increased by 20 percent so that the estimated shipping costs are changed. It is assumed in the fourth model that the port expenses in port of Houston and port of New Orleans are increased by 20 percent which in turn affect the estimated transportation costs. In the fifth model, estimated transportation costs are adjusted by reducing the operating costs of U.S.-flag vessels by 50 percent. For the last model, the construction costs for U.S.-flag ships are reduced by 50 percent. Those different transportation costs are calculated for the six models, which provide an opportunity to analyze the effects of the shipping cost on the least-cost shipping patterns of U.S. rice exports separately and collectively.

Finally, the results of all the models as well as actual shipping patterns for 1986 rice exports are evaluated and compared. The optimum shipping patterns and total transportation costs are determined from the development of the models, which are dependent on the transportation rates and the assumptions and restrictions of the model. There are no significant differences among the six total transportation costs with Models I, V, and VI having the same lowest transportation costs: $42,340,273.50. Model II had the highest total cost among all the models developed, $43,666,216.50.

In all the six models, 40,000 DWT, Liberia-flag vessels are most frequently used because of the size economies and cheaper foreign-flag ships. 30,000 DWT, Liberia-flag ships and 30,000 DWT, Japan-flag ships are used for a small volume of shipment. Of all the seven U.S. southern exporting regions, ports of Port Arthur and Houston are the principal exporting regions for large amounts of rice shipments. Ports of New Orleans, Mobile, Baton Rouge and Galveston are used to ship moderately. U.S.-flag vessels are not used at all. All the six optimum shipping patterns determined in the six models have similarities which are different from 1986 actual shipping patterns.

In summary, it is evident that transportation costs are very significant and do influence international trade. They are an important factor in determining the competitive positions of the U.S. rice exporters in the world market. Their effects should not be overlooked in the theory and study of international trade.

SUGGESTIONS

This study is not intended to advocate the immediate reorganization of rice exports from the U.S. southern region to world markets. The main purpose is to examine and analyze the effects of transportation costs on U.S. rice exports to foreign markets to provide some insights into the area and benefit U.S. rice producers and exporters.

The study is limited in several respects. The first is the lack of readily available data on international shipping costs. In fact, this has been one of the primary reasons for the absence of transportation costs in international trade studies. Since such data are simply unavailable, the needed shipping rates are estimated as a function of several factors. While this is not a totally unreasonable assumption, the fact remains that the shipping costs are influenced by some other factors that are not covered in this study. It is assumed that the efficiency of port utilization of each port is identical and has no impact on transportation costs. Certainly, such information will provide a more accurate estimation of shipping cost, and permit a better analysis of transportation costs and their effects on international trade.

Another limitation is that only major importing countries and exporting ports are included for the analysis. It is assumed that other countries and regions did not have any influence on international rice trade. While the inclusion of all the trading countries and regions could have provided a more comprehensive analysis of international rice trade, it could have made the study more complex and unmanageable and in turn obscured the original objectives.

The third limitation arises from the fact that even though there are some other factors affecting international rice trade, the study is limited only to the analysis of transportation costs. Therefore, additional research should be conducted to examine the nature and determination of international trade.

The fourth limitation is that the volume of rice shipped from the U.S. southern region and received by world market areas is assumed to be the amounts that were shipped during 1986. This data should be updated when

Summary and Conclusions

the information becomes available.

The fifth limitation is the lack of information on the size of the vessels used in actual shipments of 1986 rice exports from the U.S. southern region to world markets. This has, in turn, limited the ability to compare the actual shipping patterns with the results of the six models in terms of ship size.

This study does not take into account the cost of transportation from various production centers in the U.S. southern region to the exporting ports. A similar study should be made of transportation within the United States so that a rate matrix should be set up from production areas to the world consuming areas. An optimal shipping pattern could then be developed which would move rice from the field to the world markets at minimum cost.

This study could also be expanded to take into account the entire world rice market. Given that there are a number of rice producing and consuming regions in the world trading a homogeneous product and separated by transportation costs, with each region a single distinct market and regional supply-demand functions and surplus-deficit position known, models similar to the ones used in this study could be developed to describe and predict such things as commodity flows among regions, regional price differentials, and locational advantages of particular regions relative to others.

Finally, the reliability of the results of this study depends on the validity of the various assumptions made in this study and the accuracy of various data inputs. Therefore, careful considerations should be given in the interpretation of results obtained in the study.

Appendix A

Estimating Procedures For Vessel Construction Cost

This appendix provides a detailed procedure for calculating and estimating vessel construction cost.

(1) Acquisition cost[1] for 30,000 DWT, foreign-flag vessels (including Japan-flag and Liberia-flag) is $15,500,000.[2] This value is updated into 1986 value which is $13,500,000

(2) Construction costs for other sizes Japan-flag and Liberia-flag vessels are estimated by adding or subtracting $2,250,000[3] to/from $13,500,000, as follows:

25,000 DWT: $13,500,000 - $2,250,000 = $11,250,000
35,000 DWT: $13,500,000 + $2,250,000 = $15,750,000
40,000 DWT: $15,750,000 + $2,250,000 = $18,000,000

(3) Construction costs for U.S.-flag vessels of different sizes are estimated by using costs of foreign-flag vessels times coefficient 2.65.[4] Therefore, construction costs for vessels of each size and each flag are obtained as follows:

	U.S.-flag	Japan/Liberia-flag
25,000 DWT:	$29,810,000	$11,250,000
30,000 DWT:	$35,780,000	$13,500,000
35,000 DWT:	$41,740,000	$15,750,000
40,000 DWT:	$47,700,000	$18,000,000

NOTES

1. Construction cost used in the analysis is assumed to be equivalent to acquisition cost.
2. This value, $13,500,000 for 30,000 DWT foreign-flag ships was obtained from "*Evaluation of Maritime Protection Practices and Proposals on U.S. Bulk Trades,*" by Alexander Tsolakis, Department of Engineering Administration, The George Washington University, December 1985.
3. This value, $2,250,000, is the differences between each size of foreign-flag ships which was obtained through private communications.
4. This coefficient, 2.65, is obtained from "*An Assessment of Maritime Trade and Technology,*" page 73-74, by Office of Technology Assessment, Congress of the United States, 1983.

Appendix B

Estimating Procedures For Models II–VI

The purpose of this appendix is to provide detailed procedures for calculating and estimating the transportation costs for Models II through VI. The transportation rate for each of the five models is adjusted and estimated based on the estimation of transportation rate for the basic model which is presented in Chapter III.

MODEL II

This model tested the effect of changing interest rates on the optimal shipping pattern determined in Model I. In this model, surplus and deficit amounts are based on the data listed in tables 26 and 27. The transportation rates for the shipments are adjusted in that the interest rates for optimal ships determined in Model I are increased by 20 percent. The optimal ships determined in Model I are ships of 30,000 DWT with Japan-flag and ships of 40,000 DWT with Liberia-flag.

The adjusted interest rate:
= 12% + (12% x 20%) = 14.4%

Average annual interest payment
= [(0.1444 + 0.144/20)/2][T.C.C.] = (0.0756)(T.C.C.)

Average interest expense per voyage day
= (0.0756)(T.C.C.) / 350 voyage days/year = (0.00022)(T.C.C.)

Amortization expense per voyage day
= (0.975/7000 + 0.00022)(T.C.C.)
= (0.000139 + 0.00022)(T.C.C.)
= (0.000359)(T.C.C.)

Depreciation and interest expenses calculated for optimal ships:
Japan-flag, 30,000 DWT vessel: $4,847.00
Liberia-flag, 40,000 DWT vessel: $6,462.00

Total vessel ownership expenses for optimal ships:
Japan-flag, 30,000 DWT: $17,764.00
Liberia-flag, 40,000 DWT: $17,176.00

Voyage Cost: The total cost per ton for the optimal ships is calculated from the following relationships:

Japan-flag, 30,000 DWT vessel:
Cost/ton = 0.891 (days in port) + 1.193 (days at sea) + 0.061

Liberia-flag, 40,000 DWT vessel:
Cost/ton = 0.651 (days in port) + 0.873 (days at sea) + 0.061

Transportation rates for shipments with Japan-flag, 30,000 DWT vessels and Liberia-flag, 40,000 DWT vessels are estimated from the above relationships and are listed in table A.1. With this adjusted transportation costs, an optimal shipping pattern with least cost is obtained which is shown in table 30.

Appendix B

Table A.1. Calculated Transportation Rates for 30,000 DWT Japan-flag and 40,000 DWT Liberia-flag vessels (for Model II)

Destination	Flag	ORIGIN DWT	Mobile	Baton Rouge	New Orleans	Beaumont
			\multicolumn{4}{c}{(in dollars/ton)}			
1. (CAC)	Japan	30,000	7.20	7.68	7.20	11.24
	Liberia	40,000	6.05	6.31	5.96	8.26
2. (WSA)	Japan	30,000	13.40	13.88	13.40	17.93
	Liberia	40,000	10.50	10.85	10.50	13.14
3. (ESA)	Japan	30,000	19.49	20.09	19.61	24.37
	Liberia	40,000	14.95	15.39	15.04	17.86
4. (NCE)	Japan	30,000	18.66	19.25	18.78	23.53
	Liberia	40,000	14.34	14.78	14.43	17.25
5. (NSE)	Japan	30,000	19.01	19.61	19.25	23.89
	Liberia	40,000	14.60	15.04	14.78	17.51
6. (EA)	Japan	30,000	37.75	38.34	37.87	42.38
	Liberia	40,000	29.28	28.75	28.40	31.04
7. (WA)	Japan	30,000	25.93	26.65	26.29	30.81
	Liberia	40,000	19.67	20.19	19.93	22.57
8. (SSA)	Japan	30,000	40.61	41.09	40.61	45.36
	Liberia	40,000	30.40	30.75	30.40	33.13
9. (AO)	Japan	30,000	29.51	30.11	29.51	34.03
	Liberia	40,000	22.28	22.72	22.28	24.93
10. (WAF)	Japan	30,000	18.66	19.25	18.78	23.53
	Liberia	40,000	14.34	14.78	14.43	17.07
11. (NAF)	Japan	30,000	19.13	19.73	19.25	24.01
	Liberia	40,000	14.69	15.13	14.78	17.51
12. (EAF)	Japan	30,000	33.21	33.69	33.21	37.73
	Liberia	40,000	24.99	25.34	24.99	27.55
13. (SAF)	Japan	30,000	26.17	26.77	26.29	31.04
	Liberia	40,000	19.84	20.28	19.93	22.66

(continued on next page)

		ORIGIN	Galveston	Houston	Port Arthur
Destination	Flag	DWT			
			(in dollars/ton)		
1. (CAC)	Japan	30,000	6.67	8.57	7.68
	Liberia	40,000	8.16	6.96	6.31
2. (WSA)	Japan	30,000	12.87	14.78	14.00
	Liberia	40,000	10.11	11.50	10.94
3. (ESA)	Japan	30,000	19.32	21.22	20.44
	Liberia	40,000	14.83	16.21	15.65
4. (NCE)	Japan	30,000	18.48	20.50	19.61
	Liberia	40,000	14.21	15.69	15.04
5. (NSE)	Japan	30,000	18.96	20.86	20.09
	Liberia	40,000	14.56	15.95	15.39
6. (EA)	Japan	30,000	37.33	39.23	38.46
	Liberia	40,000	28.11	29.39	28.83
7. (WA)	Japan	30,000	25.76	27.90	26.89
	Liberia	40,000	19.54	21.01	20.36
8. (SSA)	Japan	30,000	40.31	42.21	41.44
	Liberia	40,000	30.19	31.58	31.01
9. (AO)	Japan	30,000	28.98	30.88	30.11
	Liberia	40,000	21.90	23.28	22.72
10. (WAF)	Japan	30,000	18.48	20.38	19.61
	Liberia	40,000	14.21	15.60	15.04
11. (NAF)	Japan	30,000	18.96	20.86	20.09
	Liberia	40,000	14.56	15.95	15.39
12. (EAF)	Japan	30,000	32.68	34.70	33.81
	Liberia	40,000	24.60	26.08	25.43
13. (SAF)	Japan	30,000	26.00	27.90	27.13
	Liberia	40,000	19.71	21.10	20.54

Note: Row headings are as follows: (1) CAC: Central America & the Caribbean; (2) WSA: Western South America; (3) ESA: Eastern South America; (4) NCE: Northwestern & Central Europe; (5) NSE: Northeastern & Southern Europe; (6) EA: Eastern Asia; (7) WA: Western Asia; (8) SSA: Southeastern & Southern Asia; (9) AO: Australia & Oceania; (10) WAF: Western Africa; (11) NAF: Northern Africa; (12) EAF: Eastern Africa; and (13) SAF: Southern Africa.

Appendix B

MODEL III

In this model, the effect of changing insurance rates on the optimal shipping pattern determined in Model I is tested. For the optimal ships selected by the basic model, the insurance rates are increased by 20 percent so that the estimated shipping costs are changed.

Adjusted insurance rate:
Japan-flag, 30,000 DWT vessel: $800 x 120% = $960.00
Liberia-flag, 40,000 DWT vessel: $700 x 120% = $840.00

Total vessel ownership expenses for optimal ships:
Japan-flag, 30,000 DWT vessel: $17,384.00
Liberia-flag, 40,000 DWT vessel: $16,596.00

Voyage cost: Total cost per ton for optimal ships is calculated from the following relationships.

Japan-flag, 30,000 DWT vessel:
Cost/ton = 0.873 (days in port) + 1.1744 (days at sea) + 0.061

Liberia-flag, 40,000 DWT vessel:
Cost/ton = 0.63 (days in port) + 0.852 (days at sea) + 0.056

Transportation rates for shipments with Japan-flag, 30,000 DWT vessels and Liberia-flag, 40,000 DWT vessels are estimated from the above relationships and are listed in table A.2. With this adjusted transportation costs, an optimal shipping pattern with least cost is obtained which is shown in table 31.

Table A.2. Calculated Transportation Rates for 30,000 DWT Japan-flag and 40,000 DWT Liberia-flag vessels (for Model III)

Destination	Flag	ORIGIN DWT	Mobile	Baton Rouge	New Orleans	Beaumont
			(in dollars/ton)			
1. (CAC)	Japan	30,000	7.08	7.55	7.08	10.99
	Liberia	40,000	5.87	6.13	5.79	8.01
2. (WSA)	Japan	30,000	13.18	13.65	13.18	17.57
	Liberia	40,000	10.22	10.56	10.22	12.78
3. (ESA)	Japan	30,000	19.17	19.76	19.29	23.91
	Liberia	40,000	14.56	14.99	14.65	17.38
4. (NCE)	Japan	30,000	18.35	18.94	18.47	23.09
	Liberia	40,000	13.97	14.39	14.05	16.79
5. (NSE)	Japan	30,000	18.70	19.29	18.94	23.44
	Liberia	40,000	14.22	14.65	14.39	17.04
6. (EA)	Japan	30,000	37.14	37.73	37.26	41.64
	Liberia	40,000	28.45	28.03	27.68	30.25
7. (WA)	Japan	30,000	25.51	26.22	25.87	30.25
	Liberia	40,000	19.16	19.68	19.42	21.99
8. (SSA)	Japan	30,000	39.96	40.43	39.96	44.58
	Liberia	40,000	29.64	29.98	29.64	32.29
9. (AO)	Japan	30,000	29.04	29.62	29.04	33.42
	Liberia	40,000	21.72	22.15	21.72	24.29
10. (WAF)	Japan	30,000	18.35	18.94	18.47	23.08
	Liberia	40,000	13.97	14.39	14.05	16.62
11. (NAF)	Japan	30,000	18.82	19.41	18.94	23.56
	Liberia	40,000	14.31	14.73	14.40	17.04
12. (EAF)	Japan	30,000	32.68	33.15	33.68	37.06
	Liberia	40,000	24.36	24.70	24.36	26.84
13. (SAF)	Japan	30,000	25.75	26.34	25.87	30.48
	Liberia	40,000	19.34	19.76	19.42	22.07

(continued on next page)

Appendix B

	ORIGIN		Galveston	Houston	Port Arthur
Destination	Flag	DWT			
			(in dollars/ton)		
1. (CAC)	Japan	30,000	6.56	8.42	7.55
	Liberia	40,000	5.41	6.76	6.13
2. (WSA)	Japan	30,000	12.66	14.53	13.77
	Liberia	40,000	9.84	11.19	10.65
3. (ESA)	Japan	30,000	19.00	20.87	20.11
	Liberia	40,000	14.44	15.79	15.25
4. (NCE)	Japan	30,000	18.18	20.17	19.42
	Liberia	40,000	13.85	15.28	14.65
5. (NSE)	Japan	30,000	18.65	20.52	19.76
	Liberia	40,000	14.19	15.53	15.00
6. (EA)	Japan	30,000	36.74	38.60	37.85
	Liberia	40,000	27.31	28.65	28.11
7. (WA)	Japan	30,000	25.35	27.33	26.45
	Liberia	40,000	19.05	20.47	19.85
8. (SSA)	Japan	30,000	39.67	41.45	40.78
	Liberia	40,000	29.44	30.78	30.24
9. (AO)	Japan	30,000	28.52	30.38	29.62
	Liberia	40,000	21.35	22.69	22.15
10. (WAF)	Japan	30,000	18.18	20.05	19.29
	Liberia	40,000	13.85	15.19	14.65
11. (NAF)	Japan	30,000	18.65	20.52	19.76
	Liberia	40,000	14.19	15.53	15.00
12. (EAF)	Japan	30,000	32.16	34.14	33.26
	Liberia	40,000	24.00	25.42	24.79
13. (SAF)	Japan	30,000	25.58	27.45	26.69
	Liberia	40,000	19.22	20.56	20.02

Note: Row headings are as follows: (1) CAC: Central America & the Caribbean; (2) WSA: Western South America; (3) ESA: Eastern South America; (4) NCE: Northwestern & Central Europe; (5) NSE: Northeastern & Southern Europe; (6) EA: Eastern Asia; (7) WA: Western Asia; (8) SSA: Southeastern & Southern Asia; (9) AO: Australia & Oceania; (10) WAF: Western Africa; (11) NAF: Northern Africa; (12) EAF: Eastern Africa; and (13) SAF: Southern Africa.

MODEL IV

This model examined the effect of changing transportation rates by changing some of the port expenses on the optimal shipping pattern determined in the basic model. In this model, port charges in ports of New Orleans and Houston are increased by 20 percent so that total transportation rates are adjusted. These costs are listed in table A.3.

The relationships for total cost per ton for each of the twelve ships in the ports of New Orleans and Houston are calculated as follows:

(1) 25,000 DWT, U.S.-flag vessel:
T.C./ton = 1.963 (days in port) + 2.323 (days at sea) + 0.073

(2) 25,000 DWT, Japan-flag vessel:
T.C./ton = 0.948 (days in port) + 1.308 (days at sea) + 0.073

(3) 25,000 DWT, Liberia-flag vessel:
T.C./ton = 0.718 (days in port) + 1.078 (days at sea) + 0.073

(4) 30,000 DWT, U.S.-flag vessel:
T.C./ton = 1.842 (days in port) + 2.14 (days at sea) + 0.0734

(5) 30,000 DWT, Japan-flag vessel:
T.C./ton = 0.869 (days in port) + 1.167 (days at sea) + 0.0734

(6) 30,000 DWT, Liberia-flag vessel:
T.C./ton = 0.677 (days in port) + 0.975 (days at sea) + 0.0734

(7) 35,000 DWT, U.S.-flag vessel:
T.C./ton = 1.756 (days in port) + 2.009 (days at sea) + 0.066

(8) 35,000 DWT, Japan-flag vessel:
T.C./ton = 0.8134 (days in port) + 1.066 (days at sea) + 0.066

(9) 35,000 DWT, Liberia-flag vessel:
T.C./ton = 0.649 (days in port) + 0.902 (days at sea) + 0.066

Appendix B

(10) 40,000 DWT, U.S.-flag vessel:
T.C./ton = 1.692 (days in port) + 1.91 (days at sea) + 0.067

(11) 40,000 DWT, Japan-flag vessel:
T.C./ton = 0.772 (days in port) + 0.991 (days at sea) + 0.067

(12) 40,000 DWT, Liberia-flag vessel:
T.C./ton = 0.628 (days in port) + 0.847 (days at sea) + 0.067

Transportation rates for shipment with ships of different flags and sizes from ports of New Orleans and Houston to the selected destination ports are calculated from the above relationships and are listed in tables A.4. and A.5. With this adjusted transportation costs, an optimal shipping pattern with least cost is obtained for the shipments and is shown in table 32.

Table A.3. Adjusted Port Expenses for Port of Houston and Port of New Orleans (for Model IV)

VESSEL SIZE	25,000 DWT			30,000 DWT		
VESSEL FLAG	U.S.	Japan	Liberia	U.S.	Japan	Liberia
			(in dollars)			
Expense per day	420	420	420	456	456	456
Expense per call	1,236	1,236	1,236	1,494	1,494	1,494

VESSEL SIZE	35,000 DWT			40,000 DWT		
VESSEL FLAG	U.S.	Japan	Liberia	U.S.	Japan	Liberia
			(in dollars)			
Expense per day	528	528	528	588	588	588
Expense per call	1,560	1,560	1,560	1,824	1,824	1,824

Table A.4. Calculated Transportation Rates for Shipment from Port of New Orleans to the Selected Destinations (for Model IV)

Destination SIZE FLAG	(1) CAC	(2) WSA	(3) ESA	(4) NCE	(5) NSE	(6) EA	(7) WA
				(dollars/ton)			
25,000 U.S.	14.90	26.97	39.06	37.43	38.36	74.60	52.07
Japan	7.79	14.59	21.39	20.48	21.00	41.40	28.72
Liberia	6.18	11.78	17.39	16.64	17.07	33.38	23.43
30,000 U.S.	13.86	25.00	36.12	34.62	35.47	68.86	48.10
Japan	7.05	13.12	19.19	18.37	18.84	37.04	25.72
Liberia	5.71	10.78	15.85	15.16	15.55	30.76	21.31
35,000 U.S.	14.40	24.85	35.29	33.89	34.69	66.03	46.54
Japan	7.14	12.69	18.23	17.48	17.91	34.54	24.20
Liberia	5.88	10.57	15.26	14.63	14.99	29.06	20.31
40,000 U.S.	14.18	24.00	34.04	32.70	33.47	63.26	44.73
Japan	6.91	12.06	17.22	16.53	16.92	32.38	22.77
Liberia	5.77	10.18	14.58	13.99	14.33	27.54	19.32

Destination SIZE FLAG	(8) SSA	(9) AO	(10) WAF	(11) NAF	(12) EAF	(13) SAF
			(dollars/ton)			
25,000 U.S.	79.94	58.34	37.43	38.36	65.54	52.07
Japan	44.41	32.25	20.48	21.00	36.30	28.70
Liberia	36.37	26.34	16.64	17.07	29.68	23.43
30,000 U.S.	73.78	53.88	34.62	35.48	60.51	48.10
Japan	39.73	28.87	18.37	18.84	32.49	25.72
Liberia	33.01	23.94	15.16	15.55	26.96	21.31
35,000 U.S.	70.65	51.97	33.89	34.69	58.20	46.54
Japan	36.99	27.08	17.48	17.91	30.38	24.20
Liberia	31.13	22.75	14.63	14.99	25.54	20.31
40,000 U.S.	67.66	49.89	32.70	33.47	55.81	44.74
Japan	34.66	25.44	16.52	16.92	28.52	22.77
Liberia	29.49	21.61	13.99	14.33	24.24	19.32

Appendix B

Table A.5. Calculated Transportation Rates for Shipment from Port of Houston to the Selected Destinations (for Model IV)

Destination SIZE / FLAG	(1) CAC	(2) WSA	(3) ESA	(4) NCE (dollars/ton)	(5) NSE	(6) EA	(7) WA
25,000 U.S.	17.78	29.86	42.41	41.01	41.71	77.48	55.18
Japan	9.25	16.05	23.11	22.33	22.72	42.86	30.31
Liberia	7.33	12.93	18.76	18.11	18.43	35.03	24.68
30,000 U.S.	16.56	27.68	39.24	37.96	38.60	71.55	51.01
Japan	8.39	14.45	20.76	20.06	20.41	38.38	27.18
Liberia	6.88	11.85	17.11	16.53	16.82	31.83	22.47
35,000 U.S.	16.96	27.40	38.25	37.05	37.65	68.59	49.30
Japan	8.38	13.92	19.68	19.04	19.36	35.78	25.54
Liberia	6.88	11.57	16.45	15.90	16.17	30.07	21.41
40,000 U.S.	16.63	26.56	36.88	35.73	36.31	65.72	47.38
Japan	8.08	13.23	18.58	17.99	18.28	33.55	24.03
Liberia	6.74	11.15	15.72	15.21	15.46	28.51	20.38

Destination SIZE / FLAG	(8) SSA	(9) AO	(10) WAF (dollars/ton)	(11) NAF	(12) EAF	(13) SAF
25,000 U.S.	83.28	61.22	40.77	41.70	68.65	55.41
Japan	46.13	33.71	22.20	22.72	37.89	30.44
Liberia	37.73	27.49	18.00	18.43	30.94	24.79
30,000 U.S.	76.90	56.57	37.74	38.60	63.42	51.22
Japan	41.30	30.21	19.94	20.41	33.95	27.29
Liberia	34.27	25.01	16.43	16.82	28.13	22.57
35,000 U.S.	73.61	54.52	36.85	37.65	60.95	49.50
Japan	38.44	28.31	18.93	19.36	31.72	25.65
Liberia	32.32	23.75	15.81	16.17	26.64	21.50
40,000 U.S.	70.49	52.35	35.54	36.30	58.46	47.57
Japan	35.99	26.61	17.89	18.28	29.78	24.13
Liberia	30.63	22.58	15.12	15.46	25.29	20.46

MODEL V

In this model, the effect of changing transportation costs by reducing the operating cost of U.S.-flag ships on the optimal shipping pattern determined in Model I is tested.

An operating cost is any expenditure required to place the vessel in an appropriate and safe condition to carry cargo. It includes items like manning and subsistence costs, maintenance and repair, and insurance (see *Maritime Economics* by Martin Stapford, 1988). In this model, operating cost of U.S.-flag ships of different sizes have been reduced by 50 percent so that the estimated total transportation rates are adjusted.

The adjusted total ownership expenses for U.S.-flag ships are calculated and listed in table A.6. The relationships for total cost per ton for each of the different sizes U.S.-flag ships are calculated as follows:

(1) 25,000 DWT, U.S.-flag vessel
T.C./ton = 1.646 (days in port) + 2.01 (days at sea) + 0.061

(2) 30,000 DWT, U.S.-flag vessel
T.C./ton = 1.5775 (days in port) +1.879 (days at sea)+ 0.061

(3) 35,000 DWT, U.S.-flag vessel
T.C./ton = 1.529 (days in port) + 1.7855 (days at sea) + 0.055

(4) 40,000 DWT, U.S.-flag vessel
T.C./ton = 1.493 (days in port) + 1.715 (days at sea) + 0.056

Transportation costs for 1986 rice shipment with U.S.-flag ships of different sizes over selected routes are re-estimated according to the above relationships and listed in table A.7. With the adjusted shipping costs, an optimal shipping pattern for this model is obtained and shown in table 33.

Appendix B

Table A.6. Adjusted Total Vessel Ownership Expenses for U.S.-Flag Ship (for Model V)

ITEM	25,000 DWT	30,000 DWT	35,000 DWT	40,000 DWT
		(in dollars)		
Crew wages	3,350	3,350	3,350	3,350
Subsistence	150	150	150	150
Insurance	800	800	800	800
Maintenance & repair	1,000	1,000	1,000	1,000
Depreciation & interest	9,509	11,414	13,315	15,216
Stores & supplies	800	800	800	800
In-port fuel	726	726	726	726
Return on investment	11,242	13,494	15,741	17,989
Total[a]	27,577	31,734	35,882	40,031

[a] Total value per voyage day.

MODEL VI

This model examined the effect of changing transportation costs by reducing total construction cost of U.S.-flag ships on the optimal shipping pattern determined in the basic model. Total construction cost of U.S.-flag ships of different sizes are reduced by 50 percent as calculated below:

25,000 DWT vessel: T.C.C. = $14,905,000.00
30,000 DWT vessel: T.C.C. = $17,890,000.00
35,000 DWT vessel: T.C.C. = $20,870,000.00
40,000 DWT vessel: T.C.C. = $23,850,000.00

Table A.7. Calculated Transportation Rates for U.S.-flag Vessels Over Selected Routes (for Model V)

Destination	Origin DWT	(1)	(2)	(3)	(4)	(5)	(6)	(7)
				(dollars/ton)				
(1) CAC	25,000	12.68	13.48	12.68	18.41	11.83	15.12	13.48
	30,000	12.01	12.76	12.01	19.07	11.00	14.34	12.76
	35,000	12.65	13.37	12.65	18.71	11.66	14.89	13.37
	40,000	12.78	13.29	12.61	17.77	11.63	14.79	13.29
(2) WSA	25,000	23.13	23.93	23.13	29.06	22.09	25.58	24.13
	30,000	21.78	22.53	21.78	29.59	20.77	24.11	22.72
	35,000	21.94	22.65	21.94	28.34	20.94	24.18	22.83
	40,000	21.52	22.21	21.52	27.78	20.72	23.71	22.38
(3) ESA	25,000	33.38	34.39	33.58	39.92	32.94	36.43	34.99
	30,000	31.36	32.30	31.55	39.73	30.91	34.26	32.86
	35,000	31.04	31.94	31.22	38.07	30.94	33.82	32.47
	40,000	30.27	31.13	30.44	36.87	29.98	32.97	31.64
(4) NCE	25,000	31.97	32.98	32.17	38.51	31.73	35.22	33.58
	30,000	30.05	30.99	30.23	38.41	29.60	33.13	31.55
	35,000	29.79	30.69	29.97	36.64	29.51	32.75	31.22
	40,000	29.07	29.93	29.24	35.91	28.78	31.94	30.44
(5) NSE	25,000	32.58	33.58	32.98	39.32	32.54	35.83	34.38
	30,000	30.61	31.55	31.00	38.98	30.35	33.69	32.30
	35,000	30.33	31.22	30.69	36.99	30.23	34.28	31.94
	40,000	29.58	30.44	29.93	36.00	29.63	32.45	31.13
(6) EA	25,000	64.13	65.14	64.33	70.47	63.69	66.78	65.34
	30,000	60.11	61.05	60.30	67.17	59.29	62.63	61.24
	35,000	58.36	59.26	58.54	64.67	57.73	60.78	59.43
	40,000	58.23	57.37	56.68	62.76	55.87	58.86	57.54

(continued on next page)

Appendix B

		(1)	(2)	(3)	(4)	(5)	(6)	(7)
(7)	25,000	44.23	45.44	44.84	50.77	44.20	47.49	45.84
WA	30,000	41.51	42.64	42.07	48.75	41.06	44.59	43.01
	35,000	40.69	41.76	41.22	47.35	40.41	43.64	42.11
	40,000	39.53	40.56	40.05	46.30	39.24	42.40	40.90
(8)	25,000	68.96	69.76	68.96	75.30	68.72	71.81	70.36
SSA	30,000	64.62	65.37	64.62	70.93	63.98	67.33	65.94
	35,000	62.65	63.36	62.65	68.96	62.19	65.24	63.90
	40,000	60.63	61.31	60.63	66.88	60.16	63.15	61.83
(9)	25,000	50.26	51.27	50.26	56.40	49.42	52.71	51.27
AO	30,000	47.15	48.08	47.15	54.02	46.13	49.47	48.08
	35,000	46.04	46.94	46.04	52.17	45.23	48.28	46.94
	40,000	44.68	45.53	44.68	50.53	43.87	46.86	45.53
(10)	25,000	31.97	32.98	32.17	38.51	31.74	35.02	33.58
WAF	30,000	30.05	30.99	30.23	37.11	29.60	32.94	31.55
	35,000	29.80	30.69	29.97	36.28	29.51	32.57	31.22
	40,000	29.07	29.93	29.24	35.49	28.78	31.77	30.44
(11)	25,000	32.78	33.78	32.98	39.52	32.74	35.83	34.38
NAF	30,000	30.80	31.74	30.99	37.98	30.71	33.69	32.30
	35,000	30.51	31.40	30.69	36.99	30.41	33.28	31.94
	40,000	29.76	30.61	29.93	36.05	29.46	32.45	31.13
(12)	25,000	56.49	57.30	56.49	62.43	55.85	59.14	57.50
EAF	30,000	52.97	53.72	52.97	60.21	52.33	55.49	53.91
	35,000	51.58	52.29	51.58	57.53	50.76	54.00	52.47
	40,000	50.00	50.68	50.00	56.07	49.19	52.35	50.85
(13)	25,000	44.64	45.64	44.84	51.18	44.40	47.69	46.24
SAF	30,000	41.88	42.82	42.07	49.32	41.44	44.78	43.39
	35,000	41.04	41.94	41.22	47.21	40.54	43.82	42.47
	40,000	39.87	40.73	40.05	46.47	39.58	42.57	41.25

Note: Column headings are as follows: (1) Mobile; (2) Baton Rouge; (3) New Orleans; (4) Beaumont; (5) Galveston; (6) Houston; and (7) Port Arthur.

With this new construction cost for U.S.-flag ships, amortization expenses (depreciation expense plus interest expense) and return on investment expense for these ships are changed, and total vessel ownership expenses for U.S.-flag ships are in turn changed as shown in table A.8. The relationships for total cost per ton for each size U.S.-flag ships are calculated as follows:

(1) 25,000 DWT, U.S.-flag vessel:
T.C./ton = 1.347 (days in port) + 1.711 (days at sea) + 0.061
(2) 30,000 DWT, U.S.-flag vessel:
T.C./ton = 1.226 (days in port) + 1.528 (days at sea) + 0.061
(3) 35,000 DWT, U.S.-flag vessel:
T.C./ton = 1.1408 (days in port) + 1.397 (days at sea) + 0.055
(4) 40,000 DWT, U.S.-flag vessel:
T.C./ton = 1.0765 (days in port) + 1.299 (days at sea) + 0.056

Transportation costs for the shipments with U.S.-flag ships of different sizes over selected routes are re-estimated according to the above relationships and listed in table A.9. With the adjusted transportation rates, an optimal shipping pattern for this model with least cost is obtained and shown in table 34.

Table A.8. Adjusted Total Vessel Ownership Expenses for U.S.-Flag Ship (for Model VI)

ITEM	25,000 DWT	30,000 DWT	35,000 DWT	40,000 DWT
	(in dollars)			
Crew wages	6,700	6,700	6,700	6,700
Subsistence	300	300	300	300
Insurance	1,600	1,600	1,600	1,600
Maintenance & repair	2,000	2,000	2,000	2,000
Depreciation & int.	4,755	5,707	6,658	7,608
Stores & supplies	800	800	800	800
In-port fuel	726	726	726	726
Return on investment	5,621	6,747	7,871	8,995
Total[a]	22,502	24,580	26,655	28,729

[a]Total value per voyage day.

Appendix B

Table A.9. Calculated Transportation Rates for U.S.-flag Vessels Over Selected Routes (for Model VI)

Origin Destination	DWT	(1)	(2)	(3) (dollars/ton)	(4)	(5)	(6)	(7)
(1) CAC	25,000	10.58	11.27	10.58	15.30	9.75	12.61	11.27
	30,000	9.55	10.16	9.55	14.44	8.78	11.39	10.16
	35,000	9.66	10.22	9.66	14.13	8.94	11.36	10.22
	40,000	9.45	9.84	9.32	13.06	8.76	10.91	9.84
(2) WSA	25,000	19.48	20.16	19.48	24.55	18.82	21.51	20.34
	30,000	17.50	18.11	17.50	22.85	16.88	19.33	18.26
	35,000	16.93	17.48	16.93	21.17	16.34	18.62	17.62
	40,000	16.07	16.59	16.07	20.08	15.65	17.67	16.72
(3) ESA	25,000	28.21	29.06	28.38	33.12	28.06	30.75	29.58
	30,000	25.29	26.05	25.44	30.95	25.13	27.58	26.51
	35,000	24.05	24.75	24.19	28.72	24.00	26.17	25.17
	40,000	22.70	23.35	22.83	26.97	22.66	24.68	23.74
(4) NCE	25,000	27.01	27.86	27.18	32.59	26.86	29.72	28.38
	30,000	24.22	24.98	24.37	29.88	24.21	26.67	25.44
	35,000	23.07	23.77	23.21	28.72	23.05	25.33	24.19
	40,000	21.79	22.44	21.92	26.18	21.75	23.90	22.83
(5) NSE	25,000	27.52	28.38	27.86	33.10	27.54	30.24	29.06
	30,000	24.68	25.44	25.01	30.34	24.83	27.12	26.05
	35,000	23.49	24.19	23.77	28.16	23.61	25.75	24.75
	40,000	22.18	22.83	22.44	26.57	22.27	24.31	23.35
(6) EA	25,000	54.38	55.24	54.56	59.62	54.06	56.59	55.41
	30,000	48.67	49.23	48.82	54.02	48.51	50.65	49.58
	35,000	45.42	46.12	45.56	49.81	45.12	47.26	46.26
	40,000	43.87	43.22	42.70	46.71	42.27	44.32	43.35

(continued on next page)

		(1)	(2)	(3)	(4)	(5)	(6)	(7)
(7)	25,000	37.45	38.47	37.96	43.02	37.12	40.16	38.81
WA	30,000	33.54	34.46	34.00	39.20	33.54	36.00	34.76
	35,000	31.59	32.43	32.01	36.26	31.57	33.85	32.71
	40,000	29.97	30.49	30.10	34.11	29.67	31.75	30.75
(8)	25,000	58.49	59.18	58.49	63.90	58.34	60.86	59.69
SSA	30,000	52.33	52.95	52.33	57.69	52.18	54.47	53.40
	35,000	48.78	49.34	48.78	53.18	48.47	50.75	49.75
	40,000	45.70	46.21	45.70	49.83	45.52	47.46	46.60
(9)	25,000	42.58	43.43	42.58	47.64	42.08	44.61	43.43
AO	30,000	38.12	38.89	38.12	43.32	37.81	39.96	38.89
	35,000	35.78	36.48	35.78	40.03	35.34	37.34	36.48
	40,000	33.61	34.26	33.61	37.61	33.18	35.22	34.26
(10)	25,000	27.01	27.86	27.18	32.58	26.86	29.55	28.38
WAF	30,000	24.22	24.98	24.37	29.87	24.22	26.61	25.44
	35,000	23.07	23.77	23.21	27.60	23.05	25.19	24.19
	40,000	22.05	22.44	21.92	25.92	21.75	23.89	22.83
(11)	25,000	27.69	28.55	27.86	33.27	27.71	30.24	29.06
NAF	30,000	24.83	25.59	24.98	30.48	24.83	27.12	26.05
	35,000	23.63	24.33	23.77	28.15	23.61	25.75	24.75
	40,000	22.31	22.96	22.44	26.57	22.27	24.21	23.35
(12)	25,000	47.88	48.57	47.88	52.94	47.39	50.08	48.74
EAF	30,000	42.86	43.47	42.86	48.06	42.40	44.85	43.62
	35,000	40.12	40.67	40.12	44.22	39.68	41.95	40.81
	40,000	37.68	38.16	37.68	41.51	37.21	39.38	38.28
(13)	25,000	37.79	38.64	37.96	43.36	37.81	40.33	39.16
SAF	30,000	33.94	34.61	34.00	39.35	33.84	36.14	35.07
	35,000	31.87	32.57	32.01	36.40	31.85	33.99	32.99
	40,000	30.07	30.62	30.10	34.24	29.94	31.98	31.01

Note: Column headings are as follows: (1) Mobile; (2) Baton Rouge; (3) New Orleans; (4) Beaumont; (5) Galveston; (6) Houston; and (7) Port Arthur.

Bibliography

1. Adam, Gerard F. and Jere R. Behrman. *Economic Modeling of World Commodity Policy.* Lexington Books, D.C. Health and Company, Massachusetts,1978.

2. Alabama State Docks Department at Mobile, Alabama. *Rates, Charges and Regulations Applicable at the General Cargo Piers and Warehouse.* October 29, 1976.

3. Araji, A.A., Krasselt, W.A., Schermerhon, R.W. *Transportation Costs of Idaho's Beef and Beef Products Movement.* Agricultural Experiment Station, University of Idaho, 1980.

4. Baldwin, R. E. *Nontariff Distortions of International Trade,* in Baldwin and Richardson, *International Trade and Finance.* (Boston: Little, Brown, and Co., 1974).

5. Baldwin, William L. *The Thai Rice Trade as a Vertical Market Network: Structure, Performance and Policy Implications.* Economic Development and Culture Change, January, 1974.

6. Beaver, S.H. *Ships and Shipping: The Geographical Consequences of Technological Progress.* Transportation Geogaraphy. McGraw-Hill Book Company, 1974.

7. Bennathan, Esra, and A.A. Walters. *Port Pricing and Investment Policy for Developing Countries.* New York: Oxford University Press, 1979.

8. Binkley, James K. and Bruce Harrer. *Major Determinants of Ocean Freight Rates for Grains: An Econometrics Analysis.* American Journal of Agricultural Economics, 1981.

9. Board of Commissioners of Port of New Orleans. *Rules and Regulations Applying at Public Wharves*, March 12, 1987.

10. Board of Commissioners of the Port of Port Arthur. *Rates, Charges, Rules, Regulations and Services Available at Public Wharves.* February, 1978.

11. Bressler, Raymend G.Jr., and Richard A.King. *Markets, Prices and International Trade.* New York: John Wiley & Sons, Inc., 1970. Chapter 6.

12. Bulk Systems International. *Bulk Trade, Transportation and Handling, Survey.* England: McMillan House, 1980. pp. 89-116.

13. Bureau of Accounts, Interstate Commerce Commission. *Rail Carload Cost Scales.* 1974.

14. Canterbery, Roy E. and Hans Bickel. *The Green Revolution and the World Rice Market, 1967-1975.* American Journal of Agricultural Economics, May 1971.

15. Caponiti, James E., Chief, Division of Ship Operating Costs, Maritime Administration, U.S. Department of Transportation. Private Communication. October 1987, December 1988.

16. Castillo-Manuel, Paciencia. *A Spatial Equilibrium Analysis of the Impact of Cargo Preference in the World and U.S. Grain Trade.* Unpublished Ph.D. dissertation. University of Kdaho, Dec. 1980

17. Cayemberg, Glen Dale Norbert. *An Analysis of Freight Rates and Ocean Shipping of United States Grain Exports.* Unpublished thesis, Department Agricultural Economics, Iowa State University, 1969.

18. Chacholiodes, Miltiades. *Priciples of International Economics.* New York: McGraw-Hill Book Company, 1981.

19. Charnes, A. and W. W. Cooper. *The Stepping Stone Method of*

Bibliography

Explaining Linear Programming Calculations in Transportation Problems. Management Science 1: 49-69, 1954.

20. Charney, Alberta H., Nancy D. Sidhu and John F. Due. *Short Run Cost Functions for Class II Railroads.* College of Commerce and Business Administration, University of Illinois at Urbana-Champaign.

21. Congress of the United States, Office of Technology Assessment. *An Assessment of Maritime Trade and Technology,* October 1983.

22. Cosgriff, John G. *The Cost and Operations of Exempt Motor Carriers in North Dakota.* Upper Great Plains Transportation Institute, North Dakota State University. UGPTI Report, No.33, November 1981.

23. Dantzig, G.B. *Activity Analysis of Production and Allocation.* Cowles Commission Monograph 13, 1951.

24. Davis, M. LeRoy. *Cost of Shipping United States Grain Exports to Principal World Markets.* Unpublished Master Thesis, Department of Agricultural Economics, Iowa State University, 1968.

25. Drinka, Thomas P., C. Phillip Baumel, and John J. Miller. *Estimating Rail Transport Costs for Grain and Fertilizer.* Research Bulletin 1028, University of Missouri-Columbia College of Agriculture, June 1978.

26. Efferson, J. Norman. *The Production and Marketing of Rice.* Simmons Press, New Orleans, 1952.

27. El-Amir, Mohammed Ragaa Abd El-Fattah. *Location Model for the World Rice Industry.* Unpublished Ph.D. dissertation, University of California, Berkeley, 1967.

28. Ellsworth, P. T. and J.C. Laith. *The International Economy.* MacMillan Publishing Co., Inc., New York, 1975.

29. Epperson, J.E., and Tyan, H. L. *The Effects of Increased*

Transportation Cost on the International Flows of Selected Fresh Produce in Late Spring. Division of Agricultural Economics, University of Georgia, (Date is n/a).

30. Food and Agriculture Organization of the United Nations. *FAO Trade Yearbook.* Various selected issues.

31. Frankel, Ernst G., and Henry Marcus. *Ocean Transportation.* Cambridge, Mass.: The M.I.T. Press, 1973.

32. Furtan, W. H., J. G. Nagy, and G. G. Storey. *The Impact on the Canadian Rapeseed Industry from Changes in Transport and Tariff Rates.* American Journal of Agricultural Economics, (61), 1979.

33. Geraci, Vincent J. and W. Prewo. *Bilateral Trade Flows and Transport Costs.* The Review of Economics and Statistics, March 1976.

34. Glade, Edward H. *Exporting U.S. Cotton: Trends in Marketing Costs to Foreign Outlets.* USDA, FAS, Washington, D.C., CWS-12, September 1977.

35. Glaser, L. K. *Provision of the Food Security Act of 1985.* Agriculture Information Bulletin No. 498, ERS, USDA, April 1986.

36. Grant, Warren R. and Much N. Leath. *Factors Affecting Supply, Demand, and Prices of U.S. Rice.* Economics, Statistics, and Cooperatives Service, USDA, ESCS-47, March 1979.

37. Grant, Warren R., John Beach, and William Lin. *Factors Affecting Supply, Demand, and Prices of U.S. Rice.* ERS Staff Report No. AGES 840803, ERS, National Economics Division, USDA, 1984.

38. Greater Baton Rouge Port Commission. *Rates, Charges Rules and Regulations,* February 1989.

39. Harrer, B. John. *Ocean Freight Rates and Agricultural Trade.*

Unpublished thesis, Purdue University, June 1979.

40. Harwood, Joy L. *A Domestic and International Transportation Model for U.S. Soft Red Winter Wheat Methodology, Analysis, and Implications.* Department of Agricultural Economics, A.E. Res. 86-5, Cornell University, Ithaca, New York, 1986.

41. Heady, Earl O. and Wilfred Candler. *Linear Programming Methods.* Iowa State University Press, Ames, Iowa, 1958.

42. Heavor, Treavor D. and K. R. Studer. *Ship Size and Turn-Around Time: Some Empirical Evidence.* Journal of Transport Economics and Policy, (6), 1972.

43. Heckscher, E.F. *The Effect of Foreign Trade on the Distribution of Income.* Ekonomisk Tidskrift, 1919, Reprinted in H.S. Ellis and L.M. Metzler, Readings in *the Theory of International Trade,* Homewood Ill: Irwin, 1950.

44. Hitchcock, Frank. *The Distribution of a Product From Several Sources to Numerous Localities.* Journal of Mathematics and Physics, 20, 1941.

45. Holder, Shelby H.,Jr.,Dale L. Shaw, and James C. Snyder. *A System Model of the U.S. Rice Industry.* Economic Research Service, Technical Bulletin No. 1453, USDA, Nov. 1971.

46. Holder, Shelby, and Warren R. Grant. *U.S. Rice Industry.* Economics, Statistics, and cooperatives Service, USDA, 1979.

47. Houck, James P. *Elements of Agricultural Trade Policies.* Macmillan Publishing Company, 1986.

48. Jansson, Jan Owen and Dan Schneerson. *Economies of Scale of General Cargo Ships.* Review of Economics and Statistics, 60 (1978) pp. 287-293.

49. Jolly, Curtis, Lonnie L. Fielder, Jr., and Harlon Traylor. *Selected*

Factors That Affect the Market for U.S. Rice. D.A.E. Research Report No. 581. Department of Agricultural Economics and Agribusiness, Louisiana State University, June 1981.

50. Kendall, P.M.K. *A Theory of Optimum Ship Size.* Journal of Transport Economics and Policy, (6) 1972.

51. Lawrence, S.A. *International Sea Transport, The Years Ahead.* Lexington, Mass.: D.C. Heath and Company, 1972.

52. Lu, Tonathan Jung-Hui. *The Demand for United States Rice: An Economic-Geographic Analysis.* Unpublished Ph.D. dissertation, University of Washington, 1971.

53. Maritime Research, Incorporated. *Chartering Annual.* 1985, 1986.

54. Maritime Research Incorporated. Weekly published data on ship charters, New York, 1972-1986.

55. Moneta, Carmellah. *The Estimation of Transportation Costs in International Trade.* Journal of Political Economy. Volume 67, February 1959.

56. Moser, David E. and Michael W. Woolverton. *Estimating Barge Transport Costs for Grain and Fertilizer.* Research Bulletin 1028. University of Missouri-Columbia College of Agriculture.

57. National Academy of Sciences-National Research Council. *Maritime Transportation of Unitized Cargo: A Comparative Economic Analysis of Break Bulk and Unit Load Systems.* National Academy of Sciences- National Research Council Publication 745, 1959.

58. Nersesian, Roy L. *Ships and Shipping,* Tulsa, Oklahoma, 1981.

59. Ohlin, B. *Interregional and International Trade.* Cambridge, Mass.: Harvard University Press, 1933.

60. Olechowski, Andrzej and A. J. Yeats, *Hidden Preference for*

Bibliography

Developing Countries: A Note on the U.S. Import Valuation Procedure. Quarterly Review of Economics and Business, Volume 19:3, Autumn 1979.

61. Paarlberg, Philip L., Alan J. Webb, Arthur Morey, and Jerry A. Sharples. *Impacts of Policy on U.S. Agricultural Trade.* USDA, ERS, International Economics Division, December 1984.

62. Pinar, Musa. *Analysis of Ocean Transportation Costs and Tariff Barriers in International Cotton Trade.* Unpublished Ph.D. dissertation, Mississippi State University, December 1983.

63. Port of Houston Authority, Texas. *Rates, Rules, and Regulations Governing the Houston Ship Channel and the Public Owned Wharves,* August 30, 1971.

64. Port of Pascagoula Authorities, Mississippi. *Public Terminal Facilities.* April 1988.

65. Reed's Marine Distance Tables. Published by Thomas Reed Publications Limited, London, 1978.

66. Revelt, Mary Elizabeth. *Oceanborne Grain Trade: A Descriptive Analysis of Various Components of the International Shipping Industry and Their Effects on World Grain Trade.* Unpublished thesis. Purdue University, December 1980.

67. Ricardo, David. *The Principles of Political Economy and Taxation.* J.M. Dent & Sons, Ltd., London, 1948.

68. Robinson, Ross. *Size of Ship and Turn-Around Time.* Journal of Transport Economics and Policy, (12), 1978, pp. 161-178.

69. Sangsiri, Makasiri. *An Economic Analysis of Factors Affecting United States Rice Exports.* Unpublished Ph.D. dissertation. Mississippi State University, August 1983.

70. Schmitz, Andrew, Dale Sigurdson, and Otto Doering. *Domestic Farm Policy and the Grains from Trade.* American Journal of

Agricultural Economics, (68), 1986.

71. Sharp, John W. and Hugh J. McDonald. *The Impact of Vessel Size on an Optimal System of U.S. Grain Export Facilities.* Ohio Agricultural Research and Development Center. Research Bulletin, No. 1048, October 1971.

72. Smith, Adam. *The Wealth of Nations.* Modern Library Edition, pp. 424-426.

73. Southern Cooperative Series Bulletin 236. *Assembling and Transporting Cotton to Domestic Mills and Ports by Southcentral and Southeastern Shippers.* January 1979.

74. Steele, Michael J, Traffic/Intermodel Sales Manager. *Rules and Regulations Governing Dockage, Shed Hire, and Other Services and Charges Applying at the Facilities of the Galveston Wharves.* August 1987.

75. Stennis, Earl A. and Musa Pinar. *Analysis of Ocean Transportation Cost in International Cotton Marketing.* Staff Paper Series, #57. Department of Agricultural Economics, Mississippi State University, 1976.

76. Stapford, Martin. *Maritime Economics.* 1988.

77. Sturney, S. G. *Shipping Economics.* London: MacMillan Press, Ltd., 1975.

78. Temple, Barker and Sloane, Inc., Lexington, MA. *Market Assessment of U.S. Flag Bulk/Container Vessels.* Volume 1, October 1983. U.S. Department of Commerce, National Technical Information Service.

79. Uhrig, John William. *Economic Effects of Charges on Transportation Rates and Processing Capacity of Soybean Procurement by Iowa Processors.* Unpublished Ph.D. dissertation. Iowa State University, Ames, Iowa, 1965.

Bibliography

80. UNCTAD. *Relationship Between Charges in Freight Rates and Charges in Costs of Maritime Transport and the Effect on the Export Trade of Developing Countries.* TD/B/C., 4/112, United Nations, Geneva, 1973.

81. United Nations. *Freight Markets and Level and Structure of Freight Rates.* Report by the Secretariat of UNCTAD, New York, 1969.

82. U.S.D.A. *Analysis of the Effects of Cost-of-Service Transportation Rates on the U.S. Grain Marketing System.* Technical Bulletin, No. 1484.

83. U.S.D.A., Agricultural Marketing Service. *Cost of Watermelon Handling From Grower to Retailer.* Marketing Research Report, No. 1141.

84. U.S.D.A., Agriculture Cooperative Service. *Grain Exporting Economies.* Port Elevator Cost Simulations, ACS Research Report, No. 56.

85. U.S.D.A., Economic Research Service. *World Agriculture Situation and Outlook Report.* March 1989.

86. U.S.D.A. *Rice, Outlook and Situation.* Various selected Issues.

87. U.S.D.A. *U.S. Exports.* Various selected issues.

88. U.S.D.A. *Agricultural Statistics.* Various selected issues.

89. U.S. Department of Commerce, National Technical Information Service. *Development of a Standardized U.S. Flag Dry-Bulk Carrier, Phase I, Final Report,* January 1979.

90. U.S. Department of Commerce, National Technical Information Service. *Development of a Standardized U.S. Flag Dry-Bulk Carrier, Phase II, Costs and Revenues of Dry-Bulk Carriers of 35,000 DWT.* August 1980.

91. U.S. Department of Commerce, Bureau of the Census. *Unpublished Foreign Trade Statistics,* SA705/705IT, 1986. Bureau of the Census, U.S. Department of Commerce, Washington, D.D., 1986.

92. U.S. Department of Commerce, Maritime Administration, Voyage Charter Rate Service, Nos. 1-29, 1957.

93. U.S. Department of Transportation, Urban, Mass. Transportation Administration. *Cost Analysis for Social Service Agency Transportation Providers.* January 1981.

94. Winter, A. C., Maritime Administration, U.S. Department of Commerce, New York. Data on Shipping Costs, 1968.

95. Yoon, Suk-Won. *A Special Equilibrium Analysis of the Competitive Position of the Southern U.S. Rice Industry in the International Market.* Unpublished Ph.D. dissertation. Mississippi State University, May 1988.

Index

Absolute Advantage, 32
Acquisition Cost, 146
Adam Smith, 31, 32, 50
AID, 4, 12, 18, 20, 71
Bareboat Charter, 25
Base Point, 55-58, 140
Bulk Business, 10
Bulk Carrier, 173
Bulk Ship, 10, 17
C.&.F, 26
C.I.F, 7, 26, 46
Cargo Liners, 21, 22, 138
Charter, 9, 23-28, 57
Charterer, 24-28, 57
Charter Party, 23
Commodity Tariff, 21
Comparative Advantage, 32, 33
Construction Cost, 71, 72, 74, 100, 145, 146
Containership, 11, 15
Contract of Affreightment, 25
Crew Costs, 15, 17, 18, 39
David Ricardo, 32, 50
Days at Sea, viii, 57, 61-71
Deficit Region, 101, 108
Depreciation and Interest, 39, 148
Direct Subsidies, 11, 20
Engineering Approach, 54, 86
F.A.S, 26
F.I.O, 27, 28
F.O.B, 9, 26, 46
Flag of Convenience, 14
Flag of Registry, vii, 14, 109

Food Security Act, 20, 30
Free-Discharge, 27
Free-in-and-out, 27, 28
Fuel Consumption, 25, 41, 53, 72, 73, 75
Government Subsidization, 18, 19, 28, 138
Gross Terms, 27
In-port Fuel, 40, 75, 76, 79, 80, 159, 162
Labor Theory, 33
Lay Days, viii, 57, 61
LDC, 17
Least-cost Shipping Pattern, x, 8, 101
Linear Programming, 9, 99, 101, 102, 107
Liner Business, 11
Liner, x, 11, 18-22, 28
Liner Conference, 21
Liner Parcels, 22
Maintenance and Repair, 17, 40, 73, 77-79, 158
Manning, 15, 39, 43, 158
MDO, 75, 76, 80
Merchant Marine Act, 12, 20
Nation of Registry, 14
Negotiated Fixtures, 23
OECD, 17
Open Market Fixtures, 23
Open Registry, 14, 20
Operating Cost, 73, 78, 158
Optimal Shipping Pattern, ix,

112-115, 118, 121, 124, 127-129, 132
Optimal Shipping Routes, 53, 139
Per Call, 41, 80, 81, 155
Per Day, 41, 44, 80, 81, 140
Port of Destination, x, 109
Port of Origin, x, 104, 109, 140
Preference Cargoes, 12, 18-20
Return on Investment, 40, 71, 73, 76, 77, 162
Shipowner, 14, 24-28
Shipping Cost, vii, viii, xi, 18, 28, 41, 42, 53, 54, 57, 71, 72, 81, 86, 102, 109, 127, 138
Shipping Industry, vii, 3, 9, 10, 14, 30, 41, 42, 51
Ship Size, vii, 39, 41-45, 50-52, 55, 71, 109, 138, 143
Specialization, 31-33, 45, 46
Statistical, 54, 86
Stores And Supplies, 17, 40, 78, 79
Store and Supplies, 73, 77
Subsidized Fleet, 10, 23
Subsistence, 17, 39, 73, 77-79, 158, 159, 162
Surplus, 9, 34, 57, 101, 103-105, 107-109
Tanker Fleet, 11, 23
Terms of Shipment, 26, 31

Time Charter, 25
Tramps, 23
Tramp Steamers, 21, 23, 138
Transportation Models, viii, x, xi, 53, 83, 86, 101, 107, 128, 138-140
Transportation Problem, 101, 102, 105
U.S.D.A, 7, 8, 20, 29
U.S. Cargo Preference Policies, vii, 3, 18
U.S. Dry-bulk, 11-13
U.S. Effective-control Fleet, 15
U.S. Liner Industry, 11
Volume of Trade, 41, 44, 45, 50, 54
Voyage Charter, 24, 28